Sebastian Deffner

Nonequilibrium entropy production in open and closed quantum systems

Sebastian Deffner

Nonequilibrium entropy production in open and closed quantum systems

Südwestdeutscher Verlag für Hochschulschriften

Imprint
Any brand names and product names mentioned in this book are subject to trademark, brand or patent protection and are trademarks or registered trademarks of their respective holders. The use of brand names, product names, common names, trade names, product descriptions etc. even without a particular marking in this work is in no way to be construed to mean that such names may be regarded as unrestricted in respect of trademark and brand protection legislation and could thus be used by anyone.

Publisher:
Südwestdeutscher Verlag für Hochschulschriften
is a trademark of
Dodo Books Indian Ocean Ltd., member of the OmniScriptum S.R.L Publishing group
str. A.Russo 15, of. 61, Chisinau-2068, Republic of Moldova Europe
Printed at: see last page
ISBN: 978-3-8381-2685-2

Zugl. / Approved by: Augsburg, Universität Augsburg, Dissertation, 2010

Copyright © Sebastian Deffner
Copyright © 2011 Dodo Books Indian Ocean Ltd., member of the OmniScriptum S.R.L Publishing group

Nonequilibrium entropy production in open and closed quantum systems

*Litterarum radices amaras esse,
fructus iucundiores.*
(Marcus Porcius Cato)

Contents

1 **Prologue**   7
   1.1 Thermodynamics - The theory of heat and work 7
   1.2 Organization of the thesis 8

2 **Classical systems far from equilibrium**   11
   2.1 Entropy production in the linear regime 12
   2.2 Microscopic dynamics 15
     2.2.1 Langevin equation 15
     2.2.2 Fokker-Planck equation 17
   2.3 Generalizations of the second law arbitrarily far from equilibrium   21
     2.3.1 Jarzynski's work relation 22
     2.3.2 Crooks' fluctuation theorem 24
     2.3.3 Generalization to arbitrary initial states 27
   2.4 Summary 31

3 **Dynamical properties of nonequilibrium quantum systems**   33
   3.1 Geometric approach to isolated quantum systems 33
     3.1.1 Wootters' statistical distance 34
     3.1.2 Generalization to mixed states: The Bures length 39
   3.2 Measuring the distance to equilibrium 44
     3.2.1 Green-Kubo formalism 45
     3.2.2 Fidelity for Gaussian states 46
     3.2.3 The parameterized harmonic oscillator in the linear regime   47
   3.3 Minimal quantum evolution time 52
     3.3.1 Mandelstam-Tamm type bound 52
     3.3.2 Margolus-Levitin type bound 54
     3.3.3 Quantum speed limit 58
   3.4 Summary 60

Contents

4 Unitary quantum processes in thermally isolated systems   **61**
- 4.1 Thermodynamics: Work and heat in quantum mechanics 61
  - 4.1.1 Work is not an observable 62
  - 4.1.2 Fluctuation theorem for heat exchange 64
- 4.2 Generalized Clausius inequality 67
  - 4.2.1 Irreversible entropy production 68
  - 4.2.2 Lower bound for the irreversible entropy 70
  - 4.2.3 Upper estimation of the relative entropy 75
- 4.3 Maximal rate of entropy production 76
- 4.4 Illustrative example - the parameterized oscillator 78
  - 4.4.1 Lower bound on entropy production 79
  - 4.4.2 Maximal rate of entropy production 80
- 4.5 Experimental realization in cold ion traps 81
  - 4.5.1 Experimental set-up 82
  - 4.5.2 Verifying the quantum Jarzynski equality 83
  - 4.5.3 Anharmonic corrections and fluctuating electric fields .. 85
- 4.6 Summary 92

5 Thermodynamics of open quantum systems   **93**
- 5.1 Quantum Langevin equation 93
  - 5.1.1 Caldeira-Leggett model 93
  - 5.1.2 Free particle 97
  - 5.1.3 Harmonic potential 99
- 5.2 Thermodynamics in the weak coupling limit 101
  - 5.2.1 Quantum entropy production 102
  - 5.2.2 Particular processes 106
  - 5.2.3 Jarzynski type fluctuation theorem 107
- 5.3 Statistical physics of open quantum systems 111
  - 5.3.1 Markovian approximation 112
  - 5.3.2 Quantum Brownian motion 114
  - 5.3.3 Hu-Paz-Zhang master equation 117
- 5.4 Summary 118

6 Strong coupling limit - a semiclassical approach   **119**
- 6.1 Quantum Smoluchowski dynamics 119
  - 6.1.1 Reduced dynamics in path integral formulation 120

		6.1.2	Quantum strong friction regime	121
		6.1.3	Quantum Smoluchowski equation	122
		6.1.4	Quantum enhanced escape rates	124
	6.2		Quantum fluctuation theorems in the strong damping limit	127
	6.3		Experimental verification in Josephson junctions	131
		6.3.1	RCSJ-model	132
		6.3.2	I-V characteristics	135
		6.3.3	Possible measurement procedure	140
	6.4		Summary	140

7 Epilogue — 143

A Quantum information theory — 145
- A.1 Relative entropy . . . 145
  - A.1.1 Inequalities in information theory . . . 146
  - A.1.2 Quantum relative entropy . . . 146
- A.2 Fisher information . . . 147
  - A.2.1 Relation to Kullback-Leibler divergence . . . 148
  - A.2.2 Cramér-Rao bound . . . 148
- A.3 Bures metric . . . 149
  - A.3.1 Explicit formulas . . . 150
  - A.3.2 Quantum Fisher information . . . 150

B Solution of the parametric harmonic oscillator — 153
- B.1 The parametric harmonic oscillator . . . 153
- B.2 Method of generating functions . . . 154
- B.3 Measure of adiabaticity . . . 156
- B.4 Exact transition probabilities . . . 159

C Stochastic path integrals — 161
- C.1 Definition and basic properties . . . 161
- C.2 Onsager-Machlup functional for space dependent diffusion . . . 164

D Acknowledgments — 169

Bibliography — 171

Contents

List of figures 185

1 Prologue

1.1 Thermodynamics - The theory of heat and work

Thermodynamics is the phenomenological theory describing the energy conversion of heat and work. The Scottish physicist Lord Kelvin was the first to formulate a concise definition of thermodynamics when he stated in 1854 [Tho82]:

> Thermodynamics is the subject of the relation of heat to forces acting between contiguous parts of bodies, and the relation of heat to electrical agency.

At its origins the theory of thermodynamics was developed to understand and improve heat engines. Hence, special interest lies on the dynamical properties of energy conversion processes. However, the original theory was only able to predict the behavior of physical systems by considering their *macroscopic* state functions (such as entropy, temperature, pressure or volume).

Equilibrium and nonequilibrium processes

A system is considered to be in a stationary state, if all relaxation processes have come to an end. Moreover, thermal equilibrium is characterized as a stationary state in which all thermodynamic properties of the system of interest are time-independent. If the physical system changes very slowly, and, hence, the system is in an equilibrium state at all times, the process is considered to be *quasistatic*. All real physical processes, however, contain nonequilibrium contributions. A thermodynamic system is out of equilibrium, if the system is time-dependent or fluxes are present. Due to the importance of mass or energy fluxes at the system's boundaries, it is not possible to apply the thermodynamic limit. Especially

1 Prologue

for small system sizes it becomes necessary to describe the dynamical properties including thermal fluctuations.

Quantum thermodynamics

The modern trend of miniaturization leads to the development of smaller and smaller devices, such as nanoengines and molecular motors [CZ03, Cer09, HM09]. On these very short length scales, thermal as well as quantum fluctuations become important, and usual thermodynamic quantities, such as work and heat, acquire a stochastic nature. Moreover, in the quantum regime a completely new theory had to be invented, since classical notions of work and heat are no longer valid [LH07]. The present thesis contributes to this prevailing field by the research for analytical expressions of the nonequilibrium entropy production in open and closed quantum systems. Complementary to other publications [TH09a, TH09b, CH09] our approach deals with the reduced dynamics of the system only. We are motivated by an experimental point of view in the sense that the system under consideration can always be separated into an accessible subsystem and the environment. Since, generally, the environment can be arbitrarily large, e.g. the universe, it is usually not experimentally controllable. Hence, the present thesis is interested in the thermodynamic properties of the reduced system only. To this end, we will have to deal with methods and quantities of statistical physics, conventional thermodynamics, quantum information theory and the theory of open quantum systems.

1.2 Organization of the thesis

The scope of the present thesis is to draw a bow over a wide range of coupling strengths of a quantum system to its thermal surroundings. Therefore, we start with an introductory chapter 2, in which we summarize the main developments in recent statistical physics for classical systems arbitrarily far from equilibrium. In particular, we will briefly summarize the notion of fluctuation theorems [CM93, Jar97, Cro98, HS01, Sei05] and a couple of exemplary derivations.

Then, we will turn to quantum systems and discuss in chapter 3 the dynamical properties of isolated quantum systems. Chapter 3 presents a detailed analysis of quantum peculiarities, which will have a notable impact on the thermodynamics properties discussed in the succeeding chapter. In particular, chapter 3 focuses

1.2 Organization of the thesis

on the description and implications of the dynamics of quantum systems. To this end, we will see that a geometric approach [Rup95] and the definition of statistical distances [Woo81] capture the dynamical properties. Moreover, we will propose an appropriate measure to quantify how far from equilibrium an arbitrary process operates in terms of the time averaged Bures length [Bur68, Bur69]. The definition of the Bures length will also serve as our starting point for the derivation of the generalized Heisenberg uncertainty relation [MT45, ML98, LT09]. We will derive the minimal time that an isolated quantum system needs to evolve from one state to another.

In chapter 4 we turn to a thermodynamic discussion of isolated quantum systems. Quantum mechanical work and heat, however, are not given by the eigenvalues of Hermitian operators [LH07]. Hence, we will have to deal with the quantum probability distributions of work and heat. It will turn out that the irreversible entropy production can be written as a relative entropy [Kul78, Ume62] between the current, nonequilibrium state and the corresponding equilibrium one. This identification will lead to a generalized Clausius inequality, where we will find a sharp lower bound for the irreversible entropy production in terms of the Bures length. Further, combining the quantum speed limit from chapter 3 with the analytic expression for the entropy production we will derive the maximal rate of entropy production in an isolated quantum system. The latter is a mere quantum result and a generalized version of the Bremermann-Bekenstein bound [Bre67, Bek74, Bek81, BS90]. This bound is an upper limit on the entropy, or information, that can be contained within a given finite region of space which has a finite amount of energy. In information theory this implies that there is a maximum rate of communication along a given channel. The chapter will be completed by illustrating the rigorous results with the help of the parameterized harmonic oscillator [Def08]. The latter model is the paradigm for an experimental verification of our generalized expressions of the second law of thermodynamics in modulated, cold ion traps [SSK08].

The next chapter introduces the quantum heat bath to the earlier considerations. We will analyze a quantum system coupled to an ensemble of harmonic oscillators usually called the Caldeira-Leggett model [CL81, CL83]. In contrast to the classical case, it is still an unsolved problem how to derive general expressions of the second law of thermodynamics for a reduced quantum system with arbitrary coupling to its environment. To illustrate the difficulties we will discuss the quantum Langevin equation. Nevertheless, we will be able to derive a closed expression for

1 Prologue

the irreversible entropy production by making use of solely thermodynamic arguments. It can be shown that this entropy production is the integral version of the instantaneous rate earlier derived in [Spo78, Lin83, Bre03]. Moreover, our expression of the entropy production fulfills an integral fluctuation theorem generalizing the universal form to quantum systems [Sei08]. For the sake of completeness we present, finally, a couple of quantum master equations [Lin76, CL83, PZ92] and their range of applicability.

In a last chapter 6 we complete the discussion by turning to the strong damping regime. For high friction coefficients a semiclassical description becomes possible, where the reduced dynamics of the quantum system are described by means of a quantum Smoluchowski equation [PG01, Tu04, TM07, DL09]. Quantum effects manifest themselves as additional quantum fluctuations, and, hence, an effective diffusion coefficient. We will derive Crooks and Jarzynski type fluctuation theorems by making use of a Wiener path integral representation of the solution of the evolution equation [GG79, CJ06]. Again, we will be able to propose a physical system for the experimental verification of our analytical predictions. In the case of high damping Josephson junctions are a possible choice. To prove the applicability of the quantum Smoluchowski equation we will suggest the measurement of the I-V characteristics, before we will propose a possible measurement procedure.

In the present thesis analytical expressions for the irreversible entropy production of a quantum system undergoing arbitrary nonequilibrium processes are derived for almost all kinds of couplings to a thermal environment. We start with isolated dynamics and increase the friction coefficient to continuously reach the high damping regime in the last chapter.

For the sake of simplicity of notation we will use units where the Boltzmann constant $k_B = 1$ throughout the present thesis. Hence, $\beta = 1/T$, will synonymously denote the inverse temperature and the inverse thermal energy. Moreover, we will use the shorthand notation d_x and ∂_x for the total and partial derivative with respect to x, respectively.

2 Classical systems far from equilibrium

Thermodynamics is a phenomenological theory, whose original goal was the understanding and improving of heat engines. However, it turned out to be one of the most powerful concepts explaining physical systems from chemical reactions to black holes. Its basic structure is set on two fundamental laws: the first law of thermodynamics or law of conservation of energy, and the second law of thermodynamics or entropy law. The french engineer Sadi Carnot is considered as the inventor of thermodynamics. In is most famous publication [Car24] he proposed the first systematic treatment of work and heat. He was the first to formulate the second law explaining practical experience of the construction of heat engines. To this end, he invented particular engines working in reversible cycles. In honor of his contribution the heat engine with the highest efficiency is still called *Carnot engine*. Several years later Rudolf Clausius restated Carnot's theory in a mathematical formulation [Cla64]. The first law expresses that the change in internal energy, ΔE, of a thermodynamic system is given by the sum of work, W, performed on the system and the heat, Q, transferred from the environment,

$$\Delta E = W + Q. \tag{2.1}$$

The latter Eq. (2.1) is a reformulation of the mechanic concept of the law of conservation of energy. It was one of the major developments to recognize the heat, Q, as an energy form contributing to ΔE. The key quantity of thermodynamics, however, is given by Clausius' formulation of the second law. Like work, heat is a process dependent quantity and involves details of the change of all internal degrees of freedom. Clausius proved that there is a quantity measuring the heat, which merely depends on the initial and final state of the system, the entropy S. Furthermore, the entropy change, ΔS, for the thermodynamic system under consideration is always larger than the heat exchange with its surroundings. The second

2 Classical systems far from equilibrium

law can, then, be expressed as,

$$\Delta S \geq \beta Q, \qquad (2.2)$$

where β is the inverse temperature. The latter *Clausius inequality* (2.2) is generally valid for irreversible as well as for reversible processes and for isolated as well as for open systems. The equality sign is only reached for processes in which the system is in an equilibrium state for all times. Such *quasistatic* processes are reversible and the only ones fully describable by means of conventional thermodynamics. For nonequilibrium, irreversible processes the inequality (2.2) still holds. However, an inequality is always less informative than an equality. Hence, thermodynamics has to be extended in order to fully describe nonequilibrium situations. The purpose of the present chapter is to introduce recent results and methods of nonequilibrium thermodynamics. Furthermore, we discuss statistical approaches and generalizations of the second law. Here, we concentrate on classical systems before we move to the quantum regime in the following chapters.

2.1 Entropy production in the linear regime

Let us start with an extension of thermodynamics to the *linear regime*. Here, linear regime means that the system under consideration stays close enough to equilibrium that in a first order approximation the microscopic dynamics are locally describable by equilibrium processes. The nonequilibrium phenomena under consideration are e.g. relaxation or heat conduction. The first complete description was introduced by Lars Onsager [Ons31a, Ons31b] and further elaborated by Ilya Prigogine [Pri47] and Ryogo Kubo [Kub57]. Moreover, a lucid treatment can be found by de Groot and Mazur [dGM84].

A systematic thermodynamic scheme for the description of nonequilibrium processes must also be built on the first (2.1) and second law (2.2). To extend the formulation to irreversible processes, however, it is necessary to restate these laws in a way suitable for this purpose. In the following, we mainly concentrate on the second law, since the conservation of energy can be taken for granted even in nonequilibrium situations. First, we reformulate the Clausius inequality (2.2) with the internal entropy change, ΔS_i, as an equality,

$$\Delta S = \Delta S_i + \Delta S_e, \qquad (2.3)$$

2.1 Entropy production in the linear regime

where the external entropy change is given by the heat exchanged with the environment, $\Delta S_e = \beta Q$. Accordingly, the second law (2.2) translates into,

$$\Delta S_i \geq 0, \tag{2.4}$$

which is easily understood by considering particular processes, where the total heat exchange vanishes, e.g. for isolated systems or cyclic processes. Moreover, the internal part of the entropy change, ΔS_i, has to be zero for reversible, equilibrium transformations of the system. In the following, the latter Eq. (2.4) will be called synonymously *Clausius inequality*. The external part, ΔS_e, on the other hand, may be positive, zero or negative, depending on the interaction of the system with its surroundings. Conventional thermodynamics is concerned with the study of reversible transformations. The thermodynamic description of irreversible processes, however, is interested in the relation of the quantity ΔS_i and various irreversible phenomena, which may occur inside the system. Before calculating the irreversible entropy production in terms of the quantities which characterize the irreversible phenomena, we rewrite Eqs. (2.3) and (2.4) in terms of extensive properties, as mass and energy. For extensive properties the densities, ρ, are continuous functions of space coordinates and we, hence, write,

$$S = \int^V dV \rho s, \tag{2.5a}$$

$$d_t S_e = -\int^\Omega d\Omega \mathbf{J}_{tot}, \tag{2.5b}$$

$$d_t S_i = \int^V dV s_{ir}. \tag{2.5c}$$

By d_t we denote the total derivative with respect to time, ρ is the continuous, extensive density, s the entropy per unit mass, V the volume of the system, $\mathbf{J}_{tot}$ denotes the total entropy flow per unit area and unit time, Ω is the surface of the system, and, finally, s_{ir} is the entropy source strength or irreversible entropy production per unit volume and unit time. Thus, Eq. (2.3) can be rewritten with Eqs. (2.5) and the help of the Gauß' theorem as,

$$\int^V dV \left[\partial_t (\rho s) + \text{div} \{ \mathbf{J}_{tot} \} - s_{ir} \right] = 0, \tag{2.6}$$

2 Classical systems far from equilibrium

where ∂_t is the partial derivative with respect to time. Since Eqs. (2.3) and (2.4) hold independently of the volume, it follows that

$$\partial_t(\rho s) = -\mathrm{div}\{\mathbf{J}_{\mathrm{tot}}\} + s_{\mathrm{ir}}, \qquad (2.7\mathrm{a})$$

$$s_{\mathrm{ir}} \geq 0. \qquad (2.7\mathrm{b})$$

The latter equations are the local, mathematical formulation of the second law of thermodynamics. Furthermore, (2.7a) is a balance equation for the entropy density, ρs, with the positive source term s_{ir}. For later purpose (2.7a) can be rewritten as [dGM84],

$$\rho\, \mathrm{d}_t s = -\mathrm{div}\{\mathbf{J}\} + s_{\mathrm{ir}}, \qquad (2.8)$$

where the entropy flux $\mathbf{J}$ is the difference between the total entropy flux $\mathbf{J}_{\mathrm{tot}}$ and a convective contribution. For the derivation of Eqs. (2.7) it has been implicitly assumed that the macroscopic Eqs. (2.3) and (2.4) remain valid for infinitesimally small parts of the system. This is in agreement with the assumption that macroscopic measurements on the system are really measurements of the properties of small parts, which still contain a large number of constituting particles. Hence, it makes sense to deal with local values of fundamentally macroscopic concepts as entropy and entropy production.

The main physical concept defining the linear regime is that, although the total system is not in equilibrium, there exists within small mass elements a state of *local* equilibrium. For these small mass elements the local entropy s is functionally defined by a local formulation of the first law (2.1),

$$\mathrm{d}_t s = \frac{1}{T}\mathrm{d}_t e + \frac{p}{T}\mathrm{d}_t v - \sum_{k=3}^{n} f_k\, \mathrm{d}_t x_k. \qquad (2.9)$$

In Eq. (2.9) we introduced the temperature T, the local energy e, the volume per mass unit v, the pressure p, and further thermodynamic forces f_k according to the extensive quantities x_k per mass unit. The hypothesis of local equilibrium can only be justified by virtue of the validity of the conclusions derived from it. For particular microscopic models is can be shown that the relation (2.9) only remains valid, if the system is not *too far away* from an equilibrium state. For most familiar transport phenomena the use of (2.9) is justified. By combining Eqs. (2.8) and (2.9) we obtain an expression for the irreversible entropy production s_{ir},

$$s_{\mathrm{ir}} = \mathrm{div}\{\mathbf{J}\} + \rho\left(\frac{1}{T}\mathrm{d}_t e + \frac{p}{T}\mathrm{d}_t v - \sum_{k=3}^{n} f_k\, \mathrm{d}_t x_k\right). \qquad (2.10)$$

Hence, the irreversible entropy production is given by a total differential in terms of the extensive variables,

$$s_{\text{ir}} = \text{div}\{\mathbf{J}\} - \rho \sum_{k=1}^{n} f_k \, \mathrm{d}_t x_k. \tag{2.11}$$

Concluding, we remark that the latter expression (2.11) fully describes the nonequilibrium phenomena for processes during which the system stays close to equilibrium. The latter (2.11) is the starting point of successfully understanding nonequilibrium entropy production and led to the famous Onsager reciprocal relations [Ons31a, Ons31b]. However, far from equilibrium the hypothesis of local equilibrium breaks down and a more careful treatment of the microscopic dynamics becomes necessary.

2.2 Microscopic dynamics

2.2.1 Langevin equation

In the present section we introduce the mathematical description of the microscopic dynamics of systems coupled to a thermal environment. By deriving the fluctuation-dissipation theorem in 1905 Albert Einstein [Ein05] initiated the modern research of stochastic processes. Three years later Paul Langevin, a French physicist, proposed a very different, but likewise powerful description of Brownian motion [Lan08, LG97]. Both descriptions have been analyzed to be mathematically distinct. However, they are physically equivalent tools for the study of continuous random processes. The Langevin equation is a Newtonian equation of motion with an additional force stemming from the environment,

$$M\ddot{x} + M\gamma\dot{x} + V'(x) = \xi_t. \tag{2.12}$$

By M we denote the mass of the particle, γ is the damping coefficient and $V'(x)$ a conservative force arising from a confining potential. Thus, the left hand side of Eq. (2.12) is the conventional Newtonian equation of motion for a particle in a potential. Langevin's innovation is the external force ξ_t. It describes the randomness in a small, but open system introduced by thermal fluctuations of the environment. Hence, ξ_t is a stochastic variable, which is in the simplest version

2 Classical systems far from equilibrium

assumed to be Gaussian distributed. Usually one considers Gaussian white noise, which is characterized by a δ-correlation,

$$\langle \xi_t \rangle = 0 \tag{2.13a}$$
$$\langle \xi_t \xi_s \rangle = 2D\delta(t-s), \tag{2.13b}$$

where D is the noise strength, or diffusion coefficient. Despite its apparently simple form the Langevin equation (2.12) bears mathematical difficulties. Especially the handling of the stochastic force, ξ_t, led to the study of stochastic differential equations. For further mathematical details we refer to the mathematical literature [Ris89]. For the present purpose it is sufficient to keep in mind that the underlying Newtonian equation of motion of a Brownian particle is given by Eq. (2.12). However, to justify the expression used in the following for the diffusion coefficient D let us briefly derive the fluctuation-dissipation theorem.

Fluctuation-Dissipation theorem

To this end, we rewrite the Langevin equation (2.12) for the case of a free particle, $V(x) = 0$, in terms of the velocity $v = \dot{x}$,

$$M\dot{v} + M\gamma v = \xi_t. \tag{2.14}$$

The solution of the latter first-order differential equation (2.14) can be evaluated,

$$v_t = v_0 \exp(-\gamma t) + \frac{1}{M} \int_0^t ds\, \xi_s \exp(-\gamma(t-s)), \tag{2.15}$$

where v_0 is the initial velocity. Since the Langevin force is of vanishing mean (2.13), the averaged solution $\langle v_t \rangle$ results in,

$$\langle v_t \rangle = v_0 \exp(-\gamma t). \tag{2.16}$$

Moreover, the mean-square velocity $\langle v_t^2 \rangle$ takes the form,

$$\langle v_t^2 \rangle = v_0^2 \exp(-2\gamma t)$$
$$+ \frac{1}{M^2} \int_0^t ds_1 \int_0^t ds_2 \exp(-\gamma(t-s_1)) \exp(-\gamma(t-s_2)) \langle \xi_{s_1} \xi_{s_2} \rangle. \tag{2.17}$$

2.2 Microscopic dynamics

With the help of the correlation function (2.13) the twofold integral can be written in closed form and, thus, (2.17) becomes,

$$\left\langle v_t^2 \right\rangle = v_0^2 \exp(-2\gamma t) + \frac{D}{\gamma M^2}\left(1 - \exp(-2\gamma t)\right). \tag{2.18}$$

In the stationary state, which is reached for $t \gg 1$, the exponentials become negligible and the mean-square velocity (2.18) further simplifies to,

$$\left\langle v_t^2 \right\rangle = \frac{D}{\gamma M^2} = \frac{1}{\beta M}. \tag{2.19}$$

In the long time limit the system relaxes to equilibrium. Thus, we applied for the second equality in Eq. (2.19) the equilibrium mean-square velocity of the kinetic gas theory. Concluding, we obtain the fluctuation-dissipation theorem, which relates the external noise strength D with the internal friction γ,

$$D = \frac{M\gamma}{\beta}. \tag{2.20}$$

The latter was derived earlier by Einstein [Ein05] without knowledge of the Langevin equation (2.12). Hence, Eq. (2.20) is often synonymously called the *Einstein relation*.

2.2.2 Fokker-Planck equation

The Langevin force, ξ_t, with properties (2.13) is a stochastic quantity. Hence, the left hand side of Eq. (2.12) and, in particular, the position and velocity of the Brownian particle become stochastic, as well. Therefore, the microscopic dynamics of the Brownian particle are equivalently described by the evolution equation of the probability $\mathfrak{w}(x,v,t)$ to find a particle with a velocity in the interval $(v, v+dv)$ and at a position in the interval $(x, x+dx)$. Generally, the probability distribution $\mathfrak{w}(x,v,t)$ solves a partial differential equation, the Fokker-Planck equation, of the form,

$$\partial_t \mathfrak{w}(x,v,t) = \mathscr{F}(x,v,t)\mathfrak{w}(x,v,t), \tag{2.21}$$

where the linear operator $\mathscr{F}(x,v,t)$ is given by,

$$\begin{aligned}\mathscr{F}(x,v,t) = &-\partial_x D_1^x(x,v,t) - \partial_v D_1^v(x,v,t) + \partial_x^2 D_2^{x,x}(x,v,t) \\ &+ \partial_v^2 D_2^{v,v}(x,v,t) + \partial_x \partial_v D_2^{x,v}(x,v,t) + \partial_v \partial_x D_2^{v,x}(x,v,t).\end{aligned} \tag{2.22}$$

2 Classical systems far from equilibrium

In the following, we discuss two simple, one-dimensional examples of Fokker-Planck equations, namely the Klein-Kramers and the Smoluchowski equation.

Klein-Kramers equation

The Klein-Kramers equation is an equation of motion for distribution functions in position and velocity space, which is equivalent to the full Langevin equation (2.12). In the one-dimensional case it takes the form,

$$\partial_t w = -\partial_x (vw) + \partial_v \left(\frac{V'}{M} w + \gamma v w \right) + \frac{\gamma}{M\beta} \partial_v^2 w, \qquad (2.23)$$

where $w = w(x,v,t)$. Moreover, we replaced the diffusion coefficient with the help of the fluctuation-dissipation theorem (2.20). Then, the stationary solution of Eq. (2.23) is given by an equilibrium, Boltzmann-Gibbs distribution, $w_{eq} \propto \exp\left(-\beta/2 M v^2 - \beta V\right)$. The main advantage of the Fokker-Planck equation (2.23) is that we can compute the entropy production in a specific system directly. Let us start with a macroscopic equivalent of the balance equation (2.8),

$$\dot{S} = \beta \dot{Q} + \sigma_t. \qquad (2.24)$$

If the system is initially prepared in an equilibrium state, the entropy can be identified with the Shannon entropy [HS01, Sei05],

$$S = -\int dx \int dv\, w(x,v,t) \ln w(x,v,t). \qquad (2.25)$$

The heat flux, on the other hand, is computed by noting that the internal energy of the system is given by the mean Hamiltonian, $E = \langle H \rangle$,

$$E = \int dx \int dv\, w(x,v,t) H(x,v,t). \qquad (2.26)$$

Hence, the energy flux separates into two terms,

$$\begin{aligned} \dot{E} &= \int dx \int dv\, \dot{w} H + \int dx \int dv\, w \dot{H} \\ &= \dot{Q} + \dot{W}, \end{aligned} \qquad (2.27)$$

where we identified the change in the Hamiltonian as work, W. Moreover, the variation of the probability distribution is given by the evolution equation, and,

2.2 Microscopic dynamics

hence, governed by the coupling to the environment. Therefore, we identify the term arising from the time dependence of $w(x,v,t)$ as heat, Q. Concluding, the rate of irreversible entropy production is given by,

$$\sigma_t = -\int dx \int dv \, (\dot{w} \ln w + \dot{w} \beta H) \,. \tag{2.28}$$

The latter expression (2.28) can be written with the stationary solution, $w_{eq} \propto \exp(-\beta H)$, of Eq. (2.23) as,

$$\sigma_t = -\int dx \int dv \, \dot{w} \left(\ln w - \ln w_{eq} \right) \,, \tag{2.29}$$

which reduces for a time independent Hamiltonian, i.e. when no work is performed, to the negative time derivative of the Kullback-Leibler divergence $D(.||.)$ [Kul78] between the current state and the equilibrium distribution of the system,

$$\sigma_t = -d_t \int dx \int dv \, w \left(\ln w - \ln w_{eq} \right) = -d_t D \left(w || w_{eq} \right) \,. \tag{2.30}$$

Later, we will derive the quantum generalization of Eq. (2.30) in the weak coupling limit (cf. subsection 5.2.1). The Kullback-Leibler divergence $D(w_1||w_2)$ is a non-commutative measure of the distinction between two probability distributions w_1 and w_2. Moreover, $D(w_1||w_2) \geq 0$ with equality only for identical densities. Further mathematical properties of the Kullback-Leibler divergence are postponed to appendix A.1. The latter observation (2.29) is always true, as long as the stationary solution of the system is given by a Gibbsian. More insight into the dynamics, however, can be obtained by explicitly applying the evolution equation (2.23) to the expression for σ_t in Eq. (2.28). With the help of the normalization of w, $\int dx \int dv \, w = 1$, it is a tedious but straightforward calculation to obtain,

$$\sigma_t = \frac{\gamma}{M\beta} \int dx \int dv \, \frac{(\partial_v w + w \beta M v)^2}{w} \,. \tag{2.31}$$

In the latter equation we derived an exact formula for the rate of irreversible entropy production for systems whose dynamics are described by the Klein-Kramers equation (2.23). Thus, we rewrote Eq. (2.11) in a non-local form and generalized the expression in the sense that Eq. (2.31) is derived from microscopic dynamics. Furthermore, our result in Eq. (2.31) coincides with an earlier published version in the context of the fluctuation-dissipation theorem [DK97].

2 Classical systems far from equilibrium

Smoluchowski equation

The Klein-Kramers equation (2.23) further simplifies in the limit of high damping. For systems strongly coupled to the environment the inertial term in the Langevin equation (2.12) can be neglected. The latter is equivalent to considering the dynamics of the system for very large time scales. As follows from Eq. (2.16) the relaxation time of the velocity degrees of freedom is given by $1/\gamma$ and, thus, the inertial term in Eq. (2.23) becomes negligible for times $t \gg 1/\gamma$. For these time scales the Klein-Kramers equation reduces to the Smoluchowski equation [Dav54] in terms of the marginal $p(x,t) = \int dv\, w(x,v,t)$,

$$\partial_t p(x,t) = \frac{1}{\gamma M} \partial_x \left[V'(x,t) p(x,t) \right] + \frac{1}{\beta \gamma M} \partial_x^2 p(x,t). \tag{2.32}$$

Equivalently to the consideration for the Klein-Kramers equation (2.23) we can compute the rate of irreversible entropy production (2.28), which simplifies for dynamics described by the Smoluchowski equation (2.32) to,

$$\sigma_t = -\int dx \, (\dot{p} \ln p + \dot{p} \beta V). \tag{2.33}$$

Combining Eqs. (2.32) and (2.33) we obtain with the help of the normalization, $\int dx\, p(x,t) = 1$, after a few lines of calculation,

$$\sigma_t = \frac{1}{\beta \gamma M} \int dx \, \frac{(\partial_x p + p \beta V')^2}{p}, \tag{2.34}$$

which is the high damping limit of Eq. (2.31). Our expression (2.34) is the corrected form of the entropy production identified by Daems and Nicolis [DN99]. In Eq. (15) of [DN99] the Fisher information, $\int dx\, (\partial_x p)^2 / p$ (cf. appendix A.2), of the instantaneous probability distribution is called entropy production. However, the irreversible entropy production has to nullify in equilibrium. On the contrary the Fisher information of the Gibbs distributions, $p_{\text{eq}} \propto \exp(-\beta V)$, does not vanish, and, thus, an additional term has to be included. One easily convinces oneself that our expression (2.34) fulfills the second law by being always nonnegative and vanishing in equilibrium.

The above expressions for the irreversible entropy production (2.11), (2.31) and (2.34) are sufficient to characterize nonequilibrium phenomena. However, the

2.3 Generalizations of the second law arbitrarily far from equilibrium

physical relevance is restricted to situations where the thermodynamic entropy can be identified with the Shannon entropy. This means explicitly that only system close to equilibrium are describable. The following section is dedicated to recently proposed generalizations of the second law, which are valid arbitrarily far from equilibrium.

2.3 Generalizations of the second law arbitrarily far from equilibrium

Almost two decades ago Evans, Cohen and Morris (1993) [CM93] discovered in the context of the simulation of sheared fluids how to generalize the second law. For small systems the dynamics are governed by thermal fluctuations and, thus, the second law has to be generalized in terms of probability distributions. The fluctuation theorems relate the probability to find a negative entropy production Σ with the probability of the positive value. They take the general form,

$$\frac{P(\Sigma=-A)}{P(\Sigma=A)} = \exp(-A). \qquad (2.35)$$

The main statement is that the occurrence of negative entropy production for single realizations of a particular process is exponentially rare. The fluctuation theorem is, hence, the generalization of the second law for small systems driven arbitrarily far from equilibrium. In 1995 the fluctuation theorem (2.35) was proven rigorously for deterministic dynamics [GC95] and later generalized to stochastic Langevin dynamics [Kur98] and general Markov processes [LS99]. A Brownian particle dragged in a harmonic potential, for which the fluctuation theorem is simply derivable [vZC03], was the paradigm for the first experimental verification by Wang *et al.* [SE02]. However, Eq. (2.35) bears the disadvantage that one has to identify the entropy production in general nonequilibrium systems. More easily accessible is the nonequilibrium work relation contributed by Jarzynski in 1997 [Jar97],

$$\langle \exp(-\beta W) \rangle = \exp(-\beta \Delta F), \qquad (2.36)$$

which relates the nonequilibrium work with the equilibrium free energy difference between the initial and final state. Note that the system has to start in an equilibrium state, whereas the final state can be a arbitrarily far from equilibrium. The

2 Classical systems far from equilibrium

free energy difference is computed between the initial state and the equilibrium state into which the system would relax, if it had the possibility. The Jarzynski equality generalizes the second law in the sense that a formulation of the second law can be derived from Eq. (2.36). With the help of Jensen's inequality, $\exp(\langle x \rangle) \leq \langle \exp(x) \rangle$, we conclude,

$$\exp(-\beta \Delta F) \geq \exp(-\langle \beta W \rangle), \qquad (2.37)$$

which is equivalent to,

$$\langle W \rangle \geq \Delta F. \qquad (2.38)$$

The latter equation states that the mean work is always larger than the work performed for quasistatic, isothermal processes. Thus, the second law is a corollary of the Jarzynski equality (2.36) or Eq. (2.36) the generalization of the second law to nonequilibrium. Moreover, a first experimental verification of Eq. (2.36) was proposed by Liphardt et al. [JB02] by stretching RNA-molecules.

In the rest of the section we discuss simple derivations of the Jarzynski equality (2.36) and some generalizations. For the sake of clarity we will restrict ourselves to exemplary considerations. However, the relations are universally valid and derivable under fairly universal conditions [Jar08]. In particular the coupling between system and environment can be treated generally. Furthermore, the system under consideration is driven out of equilibrium by an external work parameter, α, with Hamiltonian $H(\alpha)$. Imagine for example a cylinder, whose volume is varied by moving the piston, or a rubber band, which is stretched. In the following we consider processes in which the work parameter is changed from an initial value, α_0, at $t = 0$, to a final value, α_1, at $t = \tau$.

2.3.1 Jarzynski's work relation

Let us first consider the special case in which the system is thermodynamically isolated from the environment, while the work parameter is varied from α_0 to α_τ. The physical situation, that we have in mind, is a small system very weakly coupled to the environment. Thus, the system equilibrates with inverse temperature β for a fixed work parameter, α. The time scale of the variation of the work parameter, however, is supposed to be much shorter than the relaxation time, $1/\gamma$ (2.16). Hence, the dynamics of the system during the variation of α can be approximated by Hamilton's equations of motion to high accuracy. Specifically, let $\zeta = (\mathbf{q}, \mathbf{p})$

2.3 Generalizations of the second law arbitrarily far from equilibrium

denote a *microstate* of the system. Thus, ζ is a point in the many-dimensional phase space, which includes all relevant coordinates to specify the microscopic configurations **q** at momenta **p**. Let $H(\zeta;\alpha)$ denote the Hamiltonian of the system and the microscopic evolution is then given by,

$$\dot{\mathbf{q}} = \partial_\mathbf{p} H, \qquad \dot{\mathbf{p}} = -\partial_\mathbf{q} H. \tag{2.39}$$

If the work parameter, α, is not varied, and, hence, the system not perturbed, it equilibrates with respect to the environment. Its according distribution for fixed α is Gibbsian,

$$p_\alpha^{eq}(\zeta) = \frac{1}{Z_\alpha} \exp(-\beta H(\zeta;\alpha)), \tag{2.40}$$

where we introduced the partition function Z_α, which is associated with the free energy corresponding to the equilibrium state,

$$Z_\alpha = \int d\zeta \exp(-\beta H(\zeta;\alpha)), \qquad \beta F_\alpha = -\ln Z_\alpha. \tag{2.41}$$

Now, we consider realizations of the process induced by varying α_t over the time interval $0 \le t \le \tau$ corresponding to a specific protocol. Due to the thermal isolation during the process the work performed, W, is the net change in the internal energy,

$$W = H(\zeta_\tau(x_0);\alpha_1) - H(\zeta_0;\alpha_0), \tag{2.42}$$

where ζ_t denotes the phase space evolution. Moreover, $\zeta_\tau(x_0)$ is the final microstate of the system under the condition that it started in x_0. In order to derive the nonequilibrium work relation (2.36) we have to compute the average of $\exp(-\beta W(\zeta_0))$ over initial conditions sampled from the equilibrium distributions at α_0,

$$\begin{aligned}
\langle \exp(-\beta W) \rangle &= \int d\zeta_0\, p_{\alpha_0}^{eq}(\zeta_0) \exp(-\beta W(\zeta_0)) & (2.43\text{a}) \\
&= \frac{1}{Z_{\alpha_0}} \int d\zeta_0 \exp(-\beta H(\zeta_\tau(x_0);\alpha_\tau)) & (2.43\text{b}) \\
&= \frac{1}{Z_{\alpha_0}} \int d\zeta_\tau \left|\frac{\partial \zeta_\tau}{\partial \zeta_0}\right|^{-1} \exp(-\beta H(\zeta_\tau;\alpha_\tau)). & (2.43\text{c})
\end{aligned}$$

In Eqs. (2.43) we substituted Eqs. (2.40) and (2.42) in the second line and changed the variables of integration from ζ_0 to $\zeta_\tau(x_0)$ in the third line. Such a change of

2 Classical systems far from equilibrium

variables is permitted by the one-to-one correspondence of initial and final microstates under Hamiltonian evolution. Further, Eq. (2.43c) simplifies by making use of Liouville's theorem, which ensures conservation of phase space volume and we arrive at,

$$\langle \exp(-\beta W) \rangle = \frac{1}{Z_{\alpha_0}} \int d\zeta_\tau \exp(-\beta H(\zeta_\tau; \alpha_\tau)) \\ = \frac{Z_{\alpha_1}}{Z_{\alpha_0}} = \exp(-\beta \Delta F). \quad (2.44)$$

It is worth mentioning that the Hamiltonian approach invoked the conservation of phase space volume. The latter derivation of the Jarzynski equality (2.44) is rather restrictive. Therefore, we discuss in the next subsection a more general approach for stochastic evolution.

2.3.2 Crooks' fluctuation theorem

Next, let us consider a stochastic approach following Crooks [Cro98, Cro99]. As before we are interested in the evolution of the system for times $0 \leq t \leq \tau$, during which the work parameter, α_t, is varied according to some protocol. The process, however, is now described as a sequence, $\zeta_0, \zeta_1, ..., \zeta_N$, of microstates visited at times $t_0, t_1, ..., t_N$ as the system evolves. For the sake of simplicity we assume the time sequence to be equally distributed, $t_n = n\tau/N$, and, implicitly, $(\zeta_N; t_N) = (\zeta_\tau; \tau)$. Moreover, the dynamics are describable with the Langevin equation (2.12) with Gaussian white noise, and, thus, we assume that the evolution is a *Markov process*: given the microstate ζ_n at time t_n, the subsequent microstate ζ_{n+1} is sampled randomly from a transition probability distribution, P, that depends merely on ζ_n, but not on the microstates visited at earlier times than t_n [vK92]. Physically, that randomness arises from the contact with the environment. The Markov assumption is an equivalent formulation for sequences of the δ-correlation of white noise (2.13). We explicitly exclude memory effects leading to dependence of the transition probability, P, on more than the last microstate.

Moreover, the transition probability to the next microstate, ζ_{n+1}, depends not only on the current microstate, ζ_n, but also on the current value of the work parameter, α_n. Now, we assume a detailed balance condition [vK92] for the ratio of a *forward* process, $P(\zeta_n \rightarrow \zeta_{n+1}; \alpha_n)$, and its time reversed twin, $P(\zeta_n \leftarrow \zeta_{n+1}; \alpha_n)$,

2.3 Generalizations of the second law arbitrarily far from equilibrium

which reads,
$$\frac{P(\zeta \to \zeta'; \alpha)}{P(\zeta \leftarrow \zeta'; \alpha)} = \frac{\exp(-\beta H(x'; \alpha))}{\exp(-\beta H(x; \alpha))}. \tag{2.45}$$

When the work parameter, α, is varied in discrete time steps from α_0 to $\alpha_N = \alpha_\tau$ as a forward process, the evolution of the system during one time step is given by a sequence,

$$\text{forward}: \quad (\zeta_n, \alpha_n) \to (\zeta_n, \alpha_{n+1}) \to (\zeta_{n+1}, \alpha_{n+1}). \tag{2.46}$$

The latter sequence (2.46) represents that first the value of the work parameter is updated and is, then, followed by a random step taken by the system. A trajectory between initial, ζ_0, and final microstate, ζ_τ, is generated by first sampling ζ_0 from the initial distribution $p_{\alpha_0}^{eq}$ (2.40) and, then, repeating the sequence (2.46) in time increments, $\delta t = \tau/N$. Trajectories of the reverse process ($\alpha_0 \leftarrow \alpha_\tau$) are analogously generated. However, the starting point is sampled from $p_{\alpha_1}^{eq}$ and the system is first taking a random step and, then, the value of the work parameter is updated,

$$\text{reversed}: \quad (\zeta_{n+1}, \alpha_{n+1}) \leftarrow (\zeta_{n+1}, \alpha_n) \leftarrow (\zeta_n, \alpha_n). \tag{2.47}$$

Consequently, the net change in internal energy of the system under consideration, $\Delta E = H(\zeta_N, \alpha_N) - H(\zeta_0, \alpha_0)$, can be written as a sum of two contributions. First, the changes in energy due to variation of the work parameter,

$$W = \sum_{n=0}^{N-1} [H(x_n; \alpha_{n+1}) - H(x_n; \alpha_n)], \tag{2.48}$$

and second, changes due to transitions between microstates in phase space,

$$Q = \sum_{n=0}^{N-1} [H(x_{n+1}; \alpha_{n+1}) - H(x_n; \alpha_{n+1})]. \tag{2.49}$$

In the latter Eqs. (2.48) and (2.49) we already used notation for work, W, and heat, Q. As argued by Crooks [Cro98] the first contribution (2.48) is given by an *internal* change in energy and the second term (2.49) stems from the interaction with the environment introducing the random steps in phase space. By applying the latter identification of work and heat the first law of thermodynamics (2.1), $\Delta E = W + Q$, is formulated in discrete time steps of the microscopic evolution of

2 Classical systems far from equilibrium

the system.

The probability to generate a trajectory starting in a particular initial state, ζ_0, is given by the product of the initial distribution and all subsequent transition probabilities,

$$P^F[\Xi] = p^{eq}_{\alpha_0}(\zeta_0) \prod_{n=0}^{N-1} P(\zeta_n \to \zeta_{n+1}; \alpha_{n+1}), \qquad (2.50)$$

where the stochastic independency of the single steps is guaranteed by the Markov assumption and $\Xi = (\zeta_0 \to ... \zeta_N)$. Now, we compare the probability of a trajectory Ξ during a forward process, $P^F[\Xi]$, with the probability of the *conjugated* path, $\Xi^\dagger = (\zeta_0 \leftarrow ... \zeta_N)$, during the reversed process, $P^R[\Xi^\dagger]$. The ratio of these probabilities reads,

$$\frac{P^F[\Xi]}{P^R[\Xi^\dagger]} = \frac{p^{eq}_{\alpha_0}(\zeta_0) \prod_{n=0}^{N-1} P\left(\zeta_n \to \zeta_{n+1}; \alpha^\xi_{n+1}\right)}{p^{eq}_{\alpha_1}(\zeta_N) \prod_{n=0}^{N-1} P\left(\zeta_n \leftarrow \zeta_{n+1}; \alpha^R_{N-1-n}\right)}. \qquad (2.51)$$

In the latter equation the sequence $\{\alpha^F_0, \alpha^F_1, ..., \alpha^F_N\}$ is the protocol for varying the external work parameter from α_0 to α_τ during the forward process. Analogously, $\{\alpha^R_0, \alpha^R_1, ..., \alpha^R_N\}$ specifies the reversed process, which is related to the forward process by,

$$\alpha^R_n = \alpha^F_{N-n}. \qquad (2.52)$$

Hence, every factor $P(\zeta \to \zeta'; \alpha)$ in the numerator of the ratio (2.51) is matched by $P(\zeta \leftarrow \zeta'; \alpha)$ in the denominator. Concluding, Eq. (2.51) reduces with the help of Eqs. (2.45), (2.48) and (2.52) to [Cro98],

$$\frac{P^F[\Xi]}{P^R[\Xi^\dagger]} = \exp\left(\beta\left(W^F[\Xi] - \Delta F\right)\right), \qquad (2.53)$$

where $W^F[\Xi]$ is the work performed on the system during the forward process. The relation between $W^F[\Xi]$ and the work performed during the reversed process, $W^R[\Xi^\dagger]$, reads by Eq. (2.48),

$$W^F[\Xi] = -W^R[\Xi^\dagger] \qquad (2.54)$$

2.3 Generalizations of the second law arbitrarily far from equilibrium

for a conjugate pair of trajectories, Ξ and $\Xi^\dagger$. The work distributions, ρ^F and ρ^R, are computed by integrating over all possible realizations, i.e. all discrete trajectories of the process,

$$\rho_F(+W) = \int d\Xi\, P^F[\Xi]\, \delta\left(W - W^F[\Xi]\right) \tag{2.55a}$$

$$\rho_R(-W) = \int d\Xi\, P^R[\Xi^\dagger]\, \delta\left(W + W^R[\Xi^\dagger]\right), \tag{2.55b}$$

where $d\Xi = d\Xi^\dagger = \prod_n dx_n$. Substituting Eq. (2.53) into Eq. (2.55a),

$$\rho_F(+W) = \exp(\beta(W - \Delta F)) \int d\Xi\, P^R[\Xi^\dagger]\, \delta\left(W + W^R[\Xi^\dagger]\right), \tag{2.56}$$

it follows the Crooks fluctuations theorem [Cro99],

$$\rho_R(-W) = \exp(-\beta(W - \Delta F))\, \rho_F(+W). \tag{2.57}$$

The theorem in Eq. (2.57) is a detailed version of the Jarzynski equality (2.36), which follows from integrating Eq. (2.57) over the forward work distribution,

$$\begin{aligned}1 &= \int dW\, \rho_R(-W) = \int dW\, \exp(-\beta(W - \Delta F))\, \rho_F(+W) \\ &= \langle \exp(-\beta(W - \Delta F)) \rangle_F\,. \end{aligned} \tag{2.58}$$

The latter nonequilibrium work relations (2.36) and (2.58) are generally valid for all kind of processes arbitrarily far from equilibrium. However, they are restricted to situations, where the system starts in a thermal equilibrium state. Thus, we consider in the following subsection the generalization to arbitrary initial states.

2.3.3 Generalization to arbitrary initial states

After the first verification [SE02] fluctuation theorems have been investigated experimentally in various nonequilibrium situations [JB02, SE04, TB05, SB06]. The canonical example is a highly damped Brownian particle in a driven potential. Due to the experimental and theoretical importance of the strongly damped regime, the overdamped Langevin equation,

$$M\gamma \dot{x} + \partial_x V(x, \alpha) = \xi_t. \tag{2.59}$$

2 Classical systems far from equilibrium

with Gaussian white noise, ξ_t in Eq. (2.13) and the equivalent Smoluchowski equation (2.32) have become important tools for the analysis of classical fluctuation theorems. In the following we use a slightly generalized form of Eq. (2.59),

$$M\gamma\dot{x} = \mathfrak{F}(x,\alpha) + \xi_t, \tag{2.60}$$

where the force $\mathfrak{F}(x,\alpha)$ may contain nonconservative contributions $f(x,\alpha)$,

$$\mathfrak{F}(x,\alpha) = -\partial_x V(x,\alpha) + f(x,\alpha). \tag{2.61}$$

As before, α denotes an externally controllable work parameter. Now, the question arises, if the notions appearing in the first and second law of thermodynamics can be consistently applied to microscopic nonequilibrium processes like dragging a colloidal particle through a viscous fluid [SE02].

Stochastic thermodynamics

Concerning the first law (2.1) Sekimoto interpreted the terms in the overdamped Langevin equation (2.59) in the sense of *stochastic energetics* or *stochastic thermodynamics* [Sek98]. To this end, we rewrite Eq. (2.59),

$$0 = -(-M\gamma\dot{x} + \xi_t)\mathrm{d}x + \partial_x V(x,\alpha)\mathrm{d}x, \tag{2.62}$$

where we separated contributions stemming from the interaction with the environment and internal variations of the system. Next, we identify the change in internal energy, $\mathrm{d}e$, for a single trajectory, x, with the variation of the potential,

$$\mathrm{d}e(x,\alpha) = \mathrm{d}V(x,\alpha) = \partial_x V(x,\alpha)\mathrm{d}x + \partial_\alpha V(x,\alpha)\mathrm{d}\alpha, \tag{2.63}$$

since we are considering overdamped dynamics. Moreover, we identify the external terms in Eq. (2.62), which are governed by the damping and the noise, as the heat exchanged with the environment, $\delta q(x) = (-M\gamma\dot{x} + \xi_t)\mathrm{d}x$. Combining Eqs. (2.62) and (2.63) we then obtain,

$$0 = -\delta q(x) + \mathrm{d}e(x,\alpha) - \partial_\alpha V(x,\alpha)\mathrm{d}\alpha, \tag{2.64}$$

which is a stochastic, microscopic expression of the first law, with the work $\delta w = \partial_\alpha V(x,\alpha)\mathrm{d}\alpha$. It is worth emphasizing that the work increment, δw, is given by

2.3 Generalizations of the second law arbitrarily far from equilibrium

the partial derivative of the potential with respect to the externally controllable work parameter, α. This is the only definition of work we will use throughout the present thesis. For the second law and, in particular, entropy a proper formulation is more subtle. Usually entropy is considered as an ensemble property measuring the disorder or information content of a system. Hence, it might be questionable, if the concept of entropy is assignable to stochastic, single trajectory formulations like (2.64). The fluctuation theorem (2.57), however, relates the probability of entropy generating trajectories to those of entropy annihilating ones. Thus, a definition of entropy on the level of single trajectories is required. The idea of a stochastic entropy was first used by Crooks [Cro99] and later elaborated by Seifert [Sei05]. For the sake of generality we formulate the Smoluchowski equation (2.32) corresponding to the general Langevin equation (2.60) including nonconservative forces,

$$\partial_t p(x,t) = -\partial_x j(x,t) = -\frac{1}{\gamma M} \partial_x \left[\mathfrak{F}(x,\alpha) p(x,t) \right] + \frac{1}{\beta \gamma M} \partial_x^2 p(x,t), \qquad (2.65)$$

where we introduced the stochastic flux $j(x,t)$. In particular, for systems with nonconservative forcing, $f(x,\alpha) \neq 0$, the stationary solution of Eq. (2.65) is not a Boltzmann-Gibbs distribution. Therefore, the following considerations are completely general with respect to the initial state of the system. Next, we continue with the observation that in equilibrium the thermodynamic entropy is given by the Shannon entropy. Thus, we are interested in the dynamics of S_t, with

$$S_t = -\int dx\, p(x,t) \ln\left(p(x,t)\right), \qquad (2.66)$$

even in nonequilibrium. In the latter Eq. (2.66) $p(x,t)$ is a solution of the Smoluchowski equation (2.65). The Shannon entropy (2.66), however, takes the form of an average of a quantity s_t, $S_t = \langle s_t \rangle$, which can be interpreted as the trajectory-dependent entropy for the particle or system,

$$s_t = -\ln\left(p(x,t)\right), \qquad (2.67)$$

where the probability $p(x,t)$ is evaluated along the stochastic trajectory x_t. Furthermore, for any given trajectory x_t the quantity s_t depends on the initial state of the system, from which x_0 is sampled. Thus, s_t contains information on the whole ensemble. Similarly to above considerations in subsection 2.2.2 we are interested

2 Classical systems far from equilibrium

in the rate of entropy change. Here, however, we concentrate on the dynamics of the stochastic quantity s_t. The time derivative of s_t reads,

$$\dot{s}_t = -\frac{\partial_t p(x,t)}{p(x,t)} - \frac{\partial_x p(x,t)}{p(x,t)} \dot{x}, \qquad (2.68)$$

which can be written in terms of the probability flux $j(x,t)$,

$$\dot{s}_t = -\frac{\partial_t p(x,t)}{p(x,t)} + \beta \gamma M \frac{j(x,t)}{p(x,t)} \dot{x} - \beta \mathfrak{F}(x,\alpha) \dot{x}. \qquad (2.69)$$

The third term in Eq. (2.69) can be related to the rate of heat dissipated in the medium [Sei05],

$$\beta \dot{q}_t = \beta \mathfrak{F}(x,\alpha) \dot{x}. \qquad (2.70)$$

Then, Eq. (2.69) can be written as a balance equation (2.24) for the total trajectory dependent entropy production,

$$\begin{aligned} \dot{s}_t^{\text{ir}} &= \beta \dot{q}_t + \dot{s}_t \\ &= -\frac{\partial_t p(x,t)}{p(x,t)} + \beta \gamma M \frac{j(x,t)}{p(x,t)} \dot{x}. \end{aligned} \qquad (2.71)$$

In Eq. (2.71) we defined a trajectory dependent irreversible entropy production, $\dot{s}_t^{\text{ir}}$, which is a microscopic formulation of the rate of irreversible entropy production, σ_t, derived in Eq. (2.34). With the microscopic formulation of irreversible entropy production the fluctuation theorem can be derived, now.

Integral fluctuation theorem

Next, the fluctuation theorem follows from a path integral analysis of the stochastic dynamics. Since we propose a more general derivation at a later point in section 6.2, we, here, merely present the result. It can be shown [Sei05] that the total irreversible entropy production, $\Delta s_{\text{ir}} = \int_0^\tau dt\, \dot{s}_t^{\text{ir}}$, obeys an integral fluctuation theorem,

$$\langle \exp(-\Delta s_{\text{ir}}) \rangle = 1, \qquad (2.72)$$

generalizing the Jarzynski equality (2.36) to arbitrary, initial states. It is worth emphasizing that the integral fluctuation theorem (2.72) is truly universal, since it holds for any kind of initial condition, any time dependence of force and potential, with (for $f = 0$) and without (for $f \neq 0$) detailed balance at fixed α, and any length of trajectory t without need for waiting for equilibration. Moreover, the Jarzynski equality (2.36) is recovered by evaluating Δs_{ir} for an initial Boltzmann-Gibbs distribution.

2.4 Summary

In the present chapter we discussed generalizations and extensions of conventional thermodynamics to various nonequilibrium situations. For classical systems, which obey a local equilibrium condition, thermodynamic methods can be extended to derive the irreversible entropy production. Further, we summarized properties of the Langevin and the Fokker-Planck equation, both describing the microscopic dynamics. Especially the Fokker-Planck equations are useful to obtain an insight into the dynamics of entropy production. Finally, we discussed generalizations of the second law to systems arbitrarily far from equilibrium. The fluctuation theorems measure, on the one hand, the exponentially small probability of negative entropy production for single realizations of a process, and on the other hand, relate the nonequilibrium work with the equilibrium free energy difference. The present chapter was merely concerned with small, but still classical systems. Naturally, the question about quantum effects arises when considering smaller and smaller system sizes. The following chapters are dedicated to various generalizations of the second law for isolated and thermally coupled quantum systems.

2 Classical systems far from equilibrium

3 Dynamical properties of nonequilibrium quantum systems

The preceding chapter introduced and discussed recent generalizations of the second law for classical systems far from thermal equilibrium. As argued earlier, fluctuations become more important for smaller and smaller systems under consideration. Hence, the natural question arises, what changes, if additional to thermal fluctuations quantum uncertainty enters the game. In order to elucidate this issue, the present chapter is dedicated to a geometric approach to driven quantum systems. For the sake of simplicity, we, here, restrict ourselves to isolated systems, and, therefore, unitary dynamics,

$$i\hbar\, \partial_t |\psi_t\rangle = H_t |\psi_t\rangle, \qquad (3.1)$$

with a possibly time-dependent Hamiltonian H_t. We start by introducing a natural distance on the space of density operators. This distance will be illustrated with an application, namely determining how far from equilibrium the linear regime is valid for a parameterized harmonic oscillator. Finally, a fundamental implication, the minimal quantum evolution time, the *quantum speed limit*, is derived.

3.1 Geometric approach to isolated quantum systems

In an earlier section 2.1 we introduced the linear regime, which is to some extent close to equilibrium. Processes not describable by means of linear response theory are usually called far from equilibrium. It would be desirable to define the expression *far* more precisely than merely not being in the linear regime. Thus,

3 Dynamical properties of nonequilibrium quantum systems

the present chapter is dedicated to a geometric approach of distinguishing quantum states. By defining a physically meaningful measure we will be able to determine the distance between nonequilibrium and equilibrium states of quantum systems. Let us start with pure states in subsection 3.1.1 and later generalize to arbitrary mixed states in subsection 3.1.2.

3.1.1 Wootters' statistical distance

The geometric approach to pure quantum states can equivalently be discussed in terms of arbitrary probability distributions. Hence, we consider a distinguishability criterion for probability distributions, first. Since Wootters definition of the statistical distance [Woo81] is the starting point of our later analysis, let us, briefly, summarize the basic concepts and the derivation of the *statistical distance*.

1-dimensional probability space

We start by considering two differently weighted coins. For only two possible outcomes the according probability space is one-dimensional. Every coin can be characterized by its probability of heads, p_1 and p_2, which we call the YES outcome. A statistical distance, ℓ, can, then, be defined as,

$$\ell(p_1, p_2) = \lim_{n \to \infty} \frac{1}{\sqrt{n}} \times [\text{maximum number of mutually distinguishable intermediate probabilities}], \quad (3.2)$$

where the mutually distinct probabilities are counted in n trials. Thus, we included a factor $1/\sqrt{n}$ to ensure that the limit exists. Now, we call two probabilities p and p' distinguishable in n trials, if

$$|p - p'| \geq \Delta p + \Delta p', \quad (3.3)$$

where Δp denotes the usual standard deviation, $\Delta p = \sqrt{p(1-p)/n}$. Substituting Eq. (3.3) in the definition (3.2) we obtain for the statistical distance,

$$\ell(p_1, p_2) = \lim_{n \to \infty} \frac{1}{\sqrt{n}} \sum_{p=p_1}^{p_2} \frac{1}{2\Delta p} = \int_{p_1}^{p_2} \frac{dp}{2\sqrt{p(p-1)}}. \quad (3.4)$$

3.1 Geometric approach to isolated quantum systems

By evaluating the integral in Eq. (3.4) [BM90a] the statistical distance (3.2) between two differently weighted coins, finally, reads,

$$\ell(p_1, p_2) = \arccos\left(\sqrt{p_1}\right) - \arccos\left(\sqrt{p_2}\right). \tag{3.5}$$

It is worth noting that the statistical distance (3.5) is not the usual Euclidean distance on probability space, which is given by $|p_1 - p_2|$. The difference originates in probabilities close to $1/2$ being more difficult to distinguish than probabilities near 0 and 1. In Fig. 3.1 a series of probabilities is plotted, which are equally spaced in the sense of the statistical distance (3.2). The curves represent the distributions of the YES outcome for each of the special probabilities of YES. The statistical distance is given by the number of curves which *fit* between two given points.

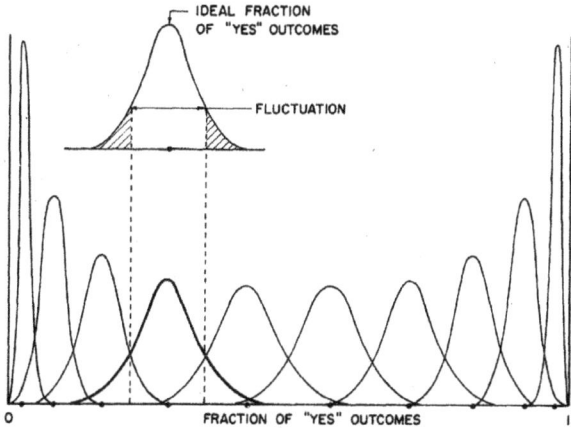

Figure 3.1: Equally spaced points in the sense of the statistical distance (3.2) (taken from [Woo81]).

d-dimensional probability space

Next, we generalize the above definition (3.2) to experiments with more than two possible outcomes. Thus, we will be able to calculate e.g. the statistical distance of

35

3 Dynamical properties of nonequilibrium quantum systems

two different non-Laplace dices or the statistical distance of two pure preparations of a general quantum system. Let us consider a probabilistic experiment with N possible outcomes. Accordingly, we have N probabilities, $p_1,..., p_N$, which span an $(N-1)$-dimensional probability space. The probability space is merely characterized by the conditions:

$$p_i \geq 0, \quad \forall \ i = 1,...,N \quad \text{and} \quad \sum_{i=1}^{N} p_i = 1. \tag{3.6}$$

Similarly to the previous case (3.2) we define the statistical distance to be proportional to the number of distinguishable points. However, in a d-dimensional space, with $d > 1$, *intermediate points* are not well-defined. Therefore, we have to refine the definition. The statistical length ℓ_C of an arbitrary curve, C, in the probability space reads,

$$\ell_C(p,p') = \lim_{n \to \infty} \frac{1}{\sqrt{n}} \times [\text{maximum number of mutually} \atop \text{distinguishable points along the curve} \, C(p \to p')], \tag{3.7}$$

where again the distinguishable probabilities are counted in n trials. The statistical distance between two points in the $(N-1)$-dimensional probability space is, then, the statistical length of the *shortest* such curve connecting two points,

$$\ell(p,p') = \min_{C(p \to p')} \ell_C(p,p'). \tag{3.8}$$

The idea of the above definition (3.8) is illustrated in Fig. 3.2 for a 3-dimensional probability space. In order to complete the definition we have to define the meaning of *distinguishable* in n trials. The actual frequencies of occurrence, $\xi_1,...,\xi_N$, for a given set of probabilities of the outcomes, $p_1,...,p_N$, are multinomially distributed. Due to the central limit theorem the multinomial distribution can be approximated by a Gaussian distribution for $n \gg 1$,

$$\rho(\xi_1,...,\xi_N) \propto \exp\left(-\frac{n}{2} \sum_{i=1}^{N} \frac{(\xi_i - p_i)^2}{p_i}\right). \tag{3.9}$$

The *region of uncertainty* around the point $p = (p_1,...,p_N)$ is defined to be the set of all points $(\xi_1,...,\xi_N)$ for which the absolute value of the exponent in Eq. (3.9)

3.1 Geometric approach to isolated quantum systems

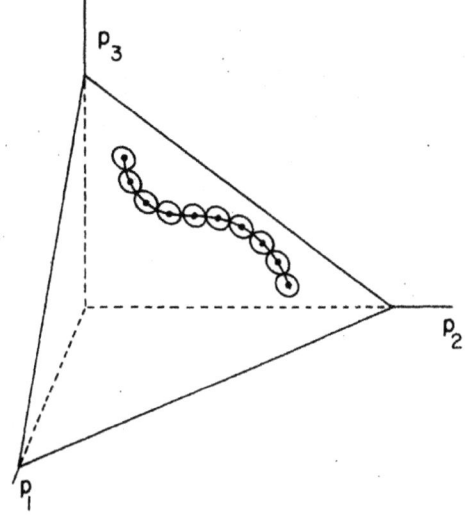

Figure 3.2: Illustration of the definition of statistical length of a path through points with regions of uncertainty in a 3-dimensional probability space (taken from [Woo81]).

is less than $1/2$. We choose $1/2$ to agree with the earlier definition of distinguishability (3.2) in the previous case, $N = 2$. Hence, two points, p and p', in the probability space are distinguishable in n trials, if their regions of uncertainty do not intersect. For $n \gg 1$, this is the case if and only if

$$\frac{\sqrt{n}}{2}\sqrt{\sum_{i=1}^{N}\frac{\delta p_i^2}{p_i}} > 1, \tag{3.10}$$

where we introduced $\delta p_i = p_i - p'_i$. The latter condition (3.10) completes the definition of the statistical distance $\ell(p, p')$ on the $(N-1)$-dimensional probability space.

Now, we want to find an explicit expression of $\ell(p, p')$, where p and p' are arbitrary points in the probability space. We parameterize the smooth curve con-

3 Dynamical properties of nonequilibrium quantum systems

necting these two points by p_t, $0 \leq t \leq \tau$ with $p_0 = p$ and $p_\tau = p'$. According to the above definition (3.7) of the statistical length and the criterion of distinguishability (3.10) we obtain,

$$\ell_C(p,p') = \frac{1}{2} \int_0^\tau dt \sqrt{\sum_{i=1}^N \frac{(d_t p_i(t))^2}{p_i(t)}}. \qquad (3.11)$$

In order to calculate the statistical distance, the minimal statistical length of a curve connecting p and p' is determined by an arbitrary optimization method. Here, we make use of a coordinate transformation, $x_i = \sqrt{p_i}$, for which Eq. (3.11) simplifies to,

$$\ell_C(p,p') = \int_0^\tau dt \sqrt{\sum_{i=1}^N (d_t x_i)^2}. \qquad (3.12)$$

Equation (3.12) is the Euclidean length of the curve in the x-space. Moreover, we rewrite the characterizing conditions of the probability space (3.6) as,

$$1 = \sum_{i=1}^N p_i(t) = \sum_{i=1}^N x_i^2(t). \qquad (3.13)$$

Now, according to Eq. (3.13) the curve $x(t)$ has to lie on the Euclidean unit sphere in the x-space. Hence, $\ell(p,p')$ is given by the shortest distance along the unit sphere between the points x and x' defined by $x = \sqrt{p}$ and $x' = \sqrt{p'}$. This shortest distance in an Euclidean space is given by the radian angle between the unit vectors x and x'. Therefore, the final expression of the statistical distance between p and p' reads,

$$\ell(p,p') = \arccos\left(\sum_{i=1}^N x_i x_i'\right) = \arccos\left(\sum_{i=1}^N \sqrt{p_i p_i'}\right). \qquad (3.14)$$

The latter result in Eq. (3.14) can be regarded as a natural notion of a distance on probability space. It takes the actual difficulty of the distinguishability of different probabilistic experiments into account.

The above definition (3.8) can be straightforwardly generalized to a probability space of countably infinite dimensions. We label the outcomes of a corresponding probabilistic experiment by $i = 1,...,\infty$. First, we want to approximate a remainder term and, thus, regard all the outcomes with $i = N+1, N+2,...$ as just one

3.1 Geometric approach to isolated quantum systems

outcome for a specific integer N. For large enough N the probability,

$$p_{N,\infty} = \sum_{i=N+1}^{\infty} p_i, \qquad (3.15)$$

of this one outcome is very small. According to Eq. (3.14) the latter approximation of the statistical distance between two points, $p = (p_1, p_2, ...)$ and $p' = (p'_1, p'_2, ...)$, in probability space reads,

$$\ell_{\text{approx}}(p, p') = \arccos\left[\left(\sum_{i=1}^{N} \sqrt{p_i p'_i}\right) + \sqrt{p_{N,\infty} p'_{N,\infty}}\right]. \qquad (3.16)$$

Finally, the exact statistical distance is defined as,

$$\ell(p, p') = \lim_{N \to \infty} \ell_{\text{approx}}(p, p') = \arccos\left(\sum_{i=1}^{\infty} \sqrt{p_i p'_i}\right). \qquad (3.17)$$

The square root appearing in Eq. (3.17) is due to Eq. (3.9) and is based on the combinatorial argument that leads to the multinomial distribution.

From the latter Eq. (3.17) we observe that the natural distinguishability distance of two pure quantum states is given by their angle in Hilbert space. In quantum mechanics the probability densities are given by the square of the wave functions' absolute values. Moreover, the argument of the $\arccos(.)$ in Eq. (3.17) is given by the overlap of two wave vectors $|\psi\rangle$ and $|\psi'\rangle$,

$$|\langle \psi | \psi' \rangle| = \sum_{i=1}^{\infty} |\langle \psi | i \rangle| \, |\langle i | \psi' \rangle| = \sum_{i=1}^{\infty} \sqrt{p_i p'_i}. \qquad (3.18)$$

Hence, generalizing Wootters' statistical distance to arbitrary, mixed quantum states is equivalent to properly defining the overlap (3.18) in the space of density operators.

3.1.2 Generalization to mixed states: The Bures length

However, for two arbitrary density operators, ρ_1 and ρ_2, it is, ad hoc, not clear at all how to determine their overlap, i.e. their fidelity function. By means of

3 Dynamical properties of nonequilibrium quantum systems

an algebraic approach Uhlmann was able to prove [Uhl76] that a properly well-defined fidelity function takes the form,

$$F(\rho_1,\rho_2) = \left[\mathrm{tr}\left\{\sqrt{\sqrt{\rho_1}\rho_2\sqrt{\rho_1}}\right\}\right]^2. \tag{3.19}$$

Uhlmann's analysis is based on general considerations of W^*-algebras following Bures [Bur68, Bur69], who derived a natural metric on the space of density operators (cf. appendix A.3). Here, we will summarize Jozsa's [Joz94] axiomatic approach in order to make the definition in Eq. (3.19) plausible. The more lucid axiomatic approach, however, bears the disadvantage that we are not able to prove the *uniqueness* of $F(\rho_1,\rho_2)$. Nevertheless, it can be shown with more abstract methods [Uhl76] that $F(\rho_1,\rho_2)$ (3.19) is, indeed, the unique definition fulfilling the following axioms and being implied by the natural generalization of Wootters' statistical distance. First, let us start by introducing the mathematical concept of *purification* for mixed quantum states in order to follow Jozsa's considerations [Joz94].

Purification of mixed quantum states

Let $\mathcal{H}_{12}$ be a Hilbert space being divided into two subspaces, $\mathcal{H}_{12} = \mathcal{H}_1 \otimes \mathcal{H}_2$, where $\otimes$ denotes the usual tensor product [BZ06]. Further, we choose the orthonormal bases $\{|e_i\rangle\}$ for $\mathcal{H}_1$ and $\{|f_j\rangle\}$ for $\mathcal{H}_2$. Then the total Hilbert space $\mathcal{H}_{12}$ is spanned by the product states $|e_i\rangle \otimes |f_j\rangle$. The basis vectors are direct products of vectors in the factor Hilbert space $\mathcal{H}_{12}$, but by taking linear combinations we will obtain vectors that cannot be written in such a form. Analogously, we can define tensor products of operators acting on single subspaces. Let A_1 act on $\mathcal{H}_1$, and A_2 acts on $\mathcal{H}_2$, then the product $A_1 \otimes A_2$ is defined by its action on the basis elements,

$$(A_1 \otimes A_2)|e_i\rangle \otimes |f_j\rangle = A_1|e_i\rangle \otimes A_2|f_j\rangle. \tag{3.20}$$

The tensor product, as defined in Eq. (3.20), is a main concept in quantum mechanics. One splits the universe into two parts, where a first part is the accessible system under consideration and a second part is the non-controllable environment. The second part may be a physical, e.g. thermal environment, or from a mathematical point of view a general device with no pretence of realism. In any case, the split is more subtle than in classical physics, since the total Hilbert space usually

3.1 Geometric approach to isolated quantum systems

contains states which cannot be written as direct products. Subsystems implying such states are called *entangled*.

Now, we take the view from the total Hilbert space $\mathcal{H}_{12}$. To compute the expectation value of an arbitrary observable we need the density matrix ρ_{12}. With the definition of the tensor product (3.20) we can, further, define the *reduced density matrices* ρ_1 and ρ_2, acting on $\mathcal{H}_1$ and $\mathcal{H}_2$, respectively, by taking *partial traces*,

$$\rho_1 = \mathrm{tr}_2\{\rho_{12}\} = \sum_{f_j} \langle e_{i'}| \otimes \langle f_j|\rho_{12}|e_i\rangle \otimes |f_j\rangle, \quad (3.21)$$

and analogously for ρ_2. The latter construction (3.21) is the mathematical formulation of the physical situation that experiments are exclusively performed on the first subsystem. In this case we are only interested in observables of the form,

$$A = A_1 \otimes \mathbb{1}_2. \quad (3.22)$$

In this case the total density operator contains more information than we need to determine the expectation value $\langle A \rangle$, since

$$\langle A \rangle = \mathrm{tr}\{\rho_{12}A\} = \mathrm{tr}_1\{\rho_1 A_1\}. \quad (3.23)$$

In the latter equation $\mathrm{tr}_1\{.\}$ denotes the trace taken over the first subsystem only. It is worth mentioning that even if ρ_{12} is a pure state, then ρ_1 will, in general, be a mixed state. Thus, we obtain mixed states by taking partial traces in larger Hilbert spaces. In the following we define the concept of purification, which turns around the latter observation. We start with a mixed state and ask for the pure state in an enlarged Hilbert space. In order to treat this property transparently, we need some further preparations in form of *Schmidt's theorem* [BZ06]:

Theorem: Every pure state $|\psi\rangle$ in the Hilbert space $\mathcal{H}_{12} = \mathcal{H}_1 \otimes \mathcal{H}_2$ can be expressed in the form

$$|\psi\rangle = \sum_i \sqrt{\lambda_i}|e_i\rangle \otimes |f_i\rangle, \quad (3.24)$$

where $\{e_i\}$ is an orthonormal basis for $\mathcal{H}_1$ and $\{f_i\}$ for $\mathcal{H}_2$.

Theorem (3.24) is also known as the *Schmidt decomposition* or *Schmidt's polar form*. The real numbers λ_i in (3.24) are called *Schmidt coefficients* with the sum rule,

$$\sum_i \lambda_i = 1, \quad \lambda_i \geq 0. \quad (3.25)$$

3 Dynamical properties of nonequilibrium quantum systems

For a proof of Schmidt's theorem (3.24) we refer to [BZ06] and, now, concentrate on two corollaries. Given any density matrix ρ on a Hilbert space $\mathcal{H}$, we can use Eq. (3.24) to find a pure state on a larger Hilbert space, whose reduction down to $\mathcal{H}$ is ρ. The first key statement can be summarized as,

> **Reduction**: Let ρ_{12} be a pure state on $\mathcal{H}_{12}$. Then the spectra of the reduced density matrices ρ_1 and ρ_2 are identical, except possibly for the degeneracy of any zero eigenvalue.

Secondly, we formulate the concept of

> **Purification**: Given a density matrix ρ_1 on a Hilbert space $\mathcal{H}_1$, there is a Hilbert space $\mathcal{H}_2$ and a pure state ρ_{12} on $\mathcal{H}_1 \otimes \mathcal{H}_2$ such that $\rho_1 = \text{tr}_2\{\rho_{12}\}$.

The above corollaries follow directly from Eq. (3.24). Their implications will become important in the following. First, purification will be used to derive Uhlmann's fidelity function (3.19) and, second, for further derivations of dynamical properties (cf. section 3.3) it suffices to consider merely pure states, without loss of generality. Generalizations to mixed states are obtained by taking partial traces over appropriately chosen subspaces of the Hilbert space under consideration.

Uhlmann's transition probability

Now, let us come back to the generalization of the fidelity to mixed states. For two pure states, $|\psi_1\rangle$ and $|\psi_1\rangle$, the natural definition of a fidelity function F (having values between 0 and 1) is provided by their overlap,

$$F(|\psi_1\rangle\langle\psi_1|, |\psi_2\rangle\langle\psi_2|) = |\langle\psi_1|\psi_2\rangle|^2 \,. \tag{3.26}$$

Equation (3.26) is in agreement with the interpretation of the statistical distance as the angle in Hilbert space (3.18). If only one state is impure then the fidelity can straightforwardly be generalized to read,

$$F(|\psi_1\rangle\langle\psi_1|, \rho_2) = \langle\psi_1|\rho_2|\psi_2\rangle \,, \tag{3.27}$$

which is the average of Eq. (3.26) over any ensemble of pure states with a density ρ_2. Now, we want to generalize Eq. (3.27) to the case of two mixed states, ρ_1 and

3.1 Geometric approach to isolated quantum systems

ρ_2. To this end, Jozsa [Joz94] formulated the following axioms to be necessary to hold for a well-defined fidelity $F(\rho_1,\rho_2)$:

F1 $0 \leq F(\rho_1,\rho_2) \leq 1$ and $F(\rho_1,\rho_2) = 1$ if and only if $\rho_1 = \rho_2$.

F2 $F(\rho_1,\rho_2) = F(\rho_2,\rho_1)$.

F3 If ρ_1 is pure, then Eq. (3.27) holds.

F4 $F(\rho_1,\rho_2)$ is invariant under unitary transformations on the state space.

From Eqs. (3.26) and (3.27) one is tempted to define $F(\rho_1,\rho_2) = \text{tr}\{\rho_1\rho_2\}$. However, for general mixed states this definition fails to satisfy F1 and it is not possible to fulfill all axioms by a simple modification. On the contrary, one easily convinces oneself that the definition in Eq. (3.19) satisfies all axioms. Moreover, Jozsa proved [Joz94] the following theorem:

Theorem: For two arbitrary mixed quantum states ρ_1 and ρ_2 the fidelity function satisfying all axioms F1-F4 reads,

$$F(\rho_1,\rho_2) = \left[\text{tr}\left\{\sqrt{\sqrt{\rho_1}\rho_2\sqrt{\rho_1}}\right\}\right]^2 = \max|\langle\phi_1|\phi_2\rangle|^2 \quad (3.28)$$

where the maximum is taken over all purifications $|\phi_1\rangle$ and $|\phi_2\rangle$ of ρ_1 and ρ_2, respectively.

In the latter theorem we assumed implicitly that the two purifications are elements of the same enlarged Hilbert space. This is justified, since without loss of generality we always can use the largest Hilbert space. For the proof of Jozsa's theorem (3.28) we refer to the literature [Joz94]. The important point is that for all theoretical analyses, i.e. not explicitly evaluating the fidelity, we can assume without loss of generality the considered states to be pure. The evaluation of $F(\rho_1,\rho_2)$ for two general densities, ρ_1 and ρ_2, is non-trivial due to the square root of operators, and, in general, only feasible for low-dimensional Hilbert spaces (cf. appendix A.3). Nevertheless, we will see an infinite dimensional example in the next section, namely $F(\rho_1,\rho_2)$ for Gaussian states. Finally, we summarize the properties of $F(\rho_1,\rho_2)$, which are given by the axioms F1-F4 and implications of (3.28):

P1 $0 \leq F(\rho_1,\rho_2) \leq 1$ and $F(\rho_1,\rho_2) = 1$ if and only if $\rho_1 = \rho_2$.

3 Dynamical properties of nonequilibrium quantum systems

P2 (Symmetry) $F(\rho_1,\rho_2) = F(\rho_2,\rho_1)$.

P3 If $\rho_1 = |\psi_1\rangle\langle\psi_1|$ is pure, then $F(\rho_1,\rho_2) = \langle\psi_1|\rho_2|\psi_1\rangle = \text{tr}\{\rho_1\rho_2\}$.

P4   a) (Convexity) If $\rho_1, \rho_2 \geq 1$ and $p_1 + p_2 = 1$, then

$$F(\rho, p_1\rho_1 + p_2\rho_2) \geq p_1 F(\rho,\rho_1) + p_2 F(\rho,\rho_2). \tag{3.29}$$

    b) $F(\rho_1,\rho_2) \geq \text{tr}\{\rho_1\rho_2\}$.

P5 (Multiplicativity) $F(\rho_1 \otimes \rho_2, \rho_3 \otimes \rho_4) = F(\rho_1,\rho_3) F(\rho_2,\rho_4)$.

P6 (Non-decreasing) $F(\rho_1,\rho_2)$ is preserved under unitary transformation. If any measurement is made on the state, transforming ρ_1, ρ_2 into ρ_1', ρ_2', then $F(\rho_1',\rho_2') \geq F(\rho_1,\rho_2)$.

Next, having properly identified the overlap of mixed states, we can generalize Wootters' statistical distance (3.18) to arbitrary quantum systems.

Generalization of Wootters' statistical distance

With the fidelity for mixed quantum states (3.19) Wootters' statistical distance (3.17) is naturally generalized by the *Bures length*,

$$\mathscr{L}(\rho_1,\rho_2) = \arccos\left(\sqrt{F(\rho_1,\rho_2)}\right) = \arccos\left(\text{tr}\left\{\sqrt{\sqrt{\rho_1}\rho_2\sqrt{\rho_1}}\right\}\right). \tag{3.30}$$

The latter definition generalizes the notion of an angle to arbitrary, mixed quantum states. The density operators are maximally distinguishable, if they are orthogonal, and the Bures length is bounded from above, $\mathscr{L} < \pi/2$. Moreover, it has been shown by Braunstein and Caves [BC94] that the underlying metric, i.e. the Bures metric (cf. appendix A.3) is the natural generalization of Wootters' infinitesimal distinguishability criterion taking explicitly non-diagonal matrix elements of the density operators into account.

3.2 Measuring the distance to equilibrium

Having defined the Bures length on the space of density operators, we come back to one of our original questions, whether we can measure the range of validity of

3.2 Measuring the distance to equilibrium

the linear regime. With the Bures length we obtained a distance between density operators and, hence, we are able to quantify how far from equilibrium a nonequilibrium process operates. For the validity of the linear regime during an arbitrary process, one supposes the system to stay close to equilibrium at all times. Therefore, the present section proposes that, generally, the time averaged Bures length has to be small for the validity of the linear regime. The time average is essential, since arbitrary processes may contain nonlinear contributions, but, nevertheless, drive the system close to an equilibrium state at single instants. To confirm the time averaged Bures length being the appropriate measure, we discuss a specific, analytically solvable example, namely the parameterized harmonic oscillator.

3.2.1 Green-Kubo formalism

Let us start with the basics concepts of the Green-Kubo formalism as a general approach to the linear regime [TH85]. To this end, we separate the Hamilton into an explicitly time-dependent contribution and the unperturbed, initial system H_0,

$$H_t = H_0 + \mathscr{V}_t, \tag{3.31}$$

where $\mathscr{V}_t$ describes a weak time-dependent perturbance. As before, we are considering a quantum process operating from $t = 0$ and to $t = \tau$, and, hence, $\mathscr{V}_t = 0$ at $t = 0^-$. The derivation simplifies by introducing the interaction picture, where the time-dependence of an operator O_t stemming from the initial Hamiltonian, H_0, is separated,

$$O_t^I = \exp(iH_0 t/\hbar) \, O_t \, \exp(-iH_0 t/\hbar). \tag{3.32}$$

Analogously, the von Neumann equation translates in the interaction picture to,

$$i\hbar \, d_t \rho_t^I = \left[\mathscr{V}_t^I, \rho_t^I \right]. \tag{3.33}$$

A solution ρ_t^I of the differential equation (3.33) equivalently solves the corresponding integral equation,

$$\rho_t^I = \rho_0 - \frac{i}{\hbar} \int_{t_0}^{t} ds \, \left[\mathscr{V}_s^I, \rho_s^I \right]. \tag{3.34}$$

The latter integral equation can be solved self-consistently by means of the Picard iteration [Aul04]. In the framework of linear response, however, the iteration

3 Dynamical properties of nonequilibrium quantum systems

takes merely the first order into account, i.e. the solution is approximated linearly. Hence, the time-dependent density operator is written as,

$$\rho_t^I \simeq \rho_0 - \frac{i}{\hbar}\int_{t_0}^{t} ds\, [\mathcal{V}_s^I, \rho_0]\,. \tag{3.35}$$

Accordingly, the mean of an arbitrary, time-dependent observable, given by $\langle O_t \rangle =$ tr $\{\rho_t^I O_t^I\}$, reads,

$$\langle O_t \rangle_{\text{linear}} = \langle O_t \rangle_{\rho_0} - \frac{i}{\hbar}\int_{t_0}^{t} ds\, \text{tr}\{\rho_0 [O_t^I, \mathcal{V}_s^I]\}\,. \tag{3.36}$$

As before, we assume the system to be initially thermalized, with density $\rho_0 = \exp(-\beta H_0)/Z_0$, and, thus we have, $[\rho_0, \exp(iH_0 t/\hbar)] = 0$. Finally, reformulating Eq. (3.36) in the Schrödinger picture the mean of a time-dependent operator is given by,

$$\langle O_t \rangle_{\text{linear}} \simeq \langle O_t \rangle_{\rho_0} - \frac{i}{\hbar}\int_{t_0}^{t} ds\, \text{tr}\{\rho_0 [O_t, \mathcal{V}_s]\}\,. \tag{3.37}$$

Equation (3.37) serves as a starting point for the derivation of the Green-Kubo formulas [TH85]. Here, however, we concentrate on applications of the Bures length (3.30). The approximations of the linear regime are valid, if the exact average of an observable equals the linear approximation, $\langle O_t \rangle_{\text{exact}} \simeq \langle O_t \rangle_{\text{linear}}$. We will see in the following that this is the case, if the Bures length averaged over the process time τ is small. First, however, we have to discuss, how the Bures length (3.30) is evaluated, i.e. how to evaluate the fidelity function (3.19).

3.2.2 Fidelity for Gaussian states

Generally, the evaluation of the fidelity function (3.19) is rather involved due to the square root of operators. However, for Gaussian, squeezed, thermal states $F(\rho_1, \rho_2)$ can be written in closed form [Scu98],

$$F(\rho_1, \rho_2) = \frac{2}{\sqrt{\Delta + \delta} - \sqrt{\delta}}, \tag{3.38}$$

3.2 Measuring the distance to equilibrium

where the parameters, Δ and δ, are defined as $\Delta = \det(A_1 + A_2)$, and the product of determinants, $\delta = (\det(A_1) - 1)(\det(A_2) - 1)$. Therefore, the fidelity for Gaussian states is merely governed by products of determinants of the according covariance matrices A_i. These are given by,

$$A_i = \begin{pmatrix} 2\left(\langle x_i^2 \rangle - \langle x_i \rangle^2\right) & \frac{2}{\hbar}\left(\frac{1}{2}\langle x_i p_i + p_i x_i \rangle - \langle x_i \rangle \langle p_i \rangle\right) \\ \frac{2}{\hbar}\left(\frac{1}{2}\langle x_i p_i + p_i x_i \rangle - \langle x_i \rangle \langle p_i \rangle\right) & \frac{2}{\hbar^2}\left(\langle p_i^2 \rangle - \langle p_i \rangle^2\right) \end{pmatrix}. \quad (3.39)$$

Consequently, the fidelity (3.38) for Gaussian states is completely determined by the first and second moments. Next, we will discuss an analytically solvable system, namely the parameterized harmonic oscillator, which is in a Gaussian state at all times. For a complete treatment of the Gaussian properties of the harmonic oscillator we refer to [Def08, AL10] and a brief review in appendix B.

3.2.3 The parameterized harmonic oscillator in the linear regime

As an illustrative example for the application of the Bures length (3.30) we choose the time-dependent Hamiltonian,

$$H_t = \frac{p^2}{2M} + \frac{M}{2}\omega_t^2 x^2. \quad (3.40)$$

where the angular frequency, ω_t, is varied from an initial value, ω_0, to a final value, ω_1, during the time interval, $0 \leq t \leq \tau$. As before, we assume the oscillator to be initially thermalized, and, hence, the initial density operator, ρ_0, reads in space representation [KP07],

$$\rho_0(x, x') = \sqrt{\frac{M\omega_0}{\pi\hbar}\tanh(\beta/2\hbar\omega_0)} \\ \times \exp\left(-\frac{M\omega_0}{2\hbar}\coth(\beta\hbar\omega_0)\left[\left(x^2 + x'^2\right) - 2\,\text{sech}(\beta\hbar\omega_0)xx'\right]\right), \quad (3.41)$$

3 Dynamical properties of nonequilibrium quantum systems

and in momentum representation,

$$\rho_0(p,p') = \sqrt{\frac{\tanh(\beta/2\hbar\omega_0)}{\pi M \hbar \omega_0}}$$
$$\times \exp\left(-\frac{1}{2M\hbar\omega_0}\coth(\beta\hbar\omega_0)\left[\left(p^2+p'^2\right) - 2\operatorname{sech}(\beta\hbar\omega_0)pp'\right]\right). \tag{3.42}$$

Let us, now, evaluate the time averaged Bures length. To this end, we ask for the distance between the instantaneous nonequilibrium state, ρ_t, and the equilibrium state corresponding to the current configuration of the system, ρ_t^{eq}. The equilibrium density operator, ρ_t^{eq}, for the current state is given by Eqs. (3.41) and (3.42) and by replacing ω_0 by ω_t everywhere. For the evaluation of the Bures length, $\mathscr{L}(\rho_t, \rho_t^{\mathrm{eq}})$, we, first, have to calculate the covariance matrices (3.39) for ρ_t and ρ_t^{eq}. The covariance matrix for the equilibrium state, ρ_t^{eq}, follows from Eqs. (3.41) and (3.42) by integration over the x- and p-space,

$$A_t^{\mathrm{eq}} = \begin{pmatrix} \frac{\hbar}{M\omega_t}\coth(\beta/2\hbar\omega_t) & 0 \\ 0 & \frac{M\omega_t}{\hbar}\coth(\beta/2\hbar\omega_t) \end{pmatrix}. \tag{3.43}$$

The instantaneous nonequilibrium density operator, $\rho_t = U_t^\dagger \rho_0 U_t$, can be written with the explicit expression for the time evolution operator (B.6) in space representation,

$$\rho_\tau(x,x') = \sqrt{\frac{M\omega_0}{\pi\hbar}\frac{\tanh(\beta/2\hbar\omega_0)}{Y_\tau^2 + \omega_0^2 X_\tau^2}}$$
$$\times \exp\left(-\frac{M\omega_0}{2\hbar}\frac{\coth(\beta\hbar\omega_0)}{Y_\tau^2 + \omega_0^2 X_\tau^2}\left[\left(x^2+x'^2\right) - 2\operatorname{sech}(\beta\hbar\omega_0)xx'\right]\right), \tag{3.44}$$

and in momentum representation

$$\rho_\tau(p,p') = \sqrt{\frac{\omega_0}{\pi M \hbar}\frac{\tanh(\beta/2\hbar\omega_0)}{\dot{Y}_\tau^2 + \omega_0^2 \dot{X}_\tau^2}}$$
$$\times \exp\left(-\frac{\omega_0}{2M\hbar}\frac{\coth(\beta\hbar\omega_0)}{\dot{Y}_\tau^2 + \omega_0^2 \dot{X}_\tau^2}\left[\left(p^2+p'^2\right) - 2\operatorname{sech}(\beta\hbar\omega_0)pp'\right]\right). \tag{3.45}$$

3.2 Measuring the distance to equilibrium

In the latter Eqs. (3.44) and (3.44) the functions X_t and Y_t are solutions of the force free classical equation (B.4) with boundary conditions $X_0 = 0$, $\dot{X}_0 = 1$ and $Y_0 = 1$, $\dot{Y}_0 = 0$. Accordingly, the covariance matrix at time t is given by,

$$A_t = \begin{pmatrix} \frac{\hbar}{M\omega_0} \left(Y_t^2 + \omega_0^2 X_t^2\right) \coth\left(\frac{\beta}{2}\hbar\omega_0\right) & \frac{1}{\omega_0} \left(Y_t \dot{Y}_t + \omega_0^2 X_t \dot{X}_t\right) \coth\left(\frac{\beta}{2}\hbar\omega_0\right) \\ \frac{1}{\omega_0} \left(Y_t \dot{Y}_t + \omega_0^2 X_t \dot{X}_t\right) \coth\left(\frac{\beta}{2}\hbar\omega_0\right) & \frac{M}{\hbar\omega_0} \left(\dot{Y}_t^2 + \omega_0^2 \dot{X}_t^2\right) \coth\left(\frac{\beta}{2}\hbar\omega_0\right) \end{pmatrix}, \tag{3.46}$$

where we used that $\langle xp + px \rangle = M\,d_t \langle x^2 \rangle$. Finally, by making use of the mathematical properties of X_t and Y_t [Def08] the fidelity function between the instantaneous nonequilibrium density operator and its equilibrium counter part results in (cf. subsection 3.2.2),

$$F\left(\rho_t, \rho_t^{eq}\right) = 2\left[-\text{ch}(\beta/2\,\varepsilon_0)\text{ch}(\beta/2\,\varepsilon_t) + \left(\coth^2(\beta/2\,\varepsilon_0) + \coth^2(\beta/2\,\varepsilon_t)\right.\right.$$
$$\left.\left. + 2Q_t^*\,\text{ch}(\beta/2\,\varepsilon_0)\text{ch}(\beta/2\,\varepsilon_t) + \text{ch}^2(\beta/2\,\varepsilon_0)\text{ch}^2(\beta/2\,\varepsilon_t)\right)^{1/2}\right]^{-1}, \tag{3.47}$$

where we introduced the energies, $\varepsilon_t = \hbar\omega_t$, and the measure of adiabaticity, Q_t^* in Eq. (B.11). Moreover, the ch(.)-function is a short notation for cosh(.). Now, the time averaged Bures length is given with Eq. (3.47) by,

$$\langle \mathscr{L} \rangle_\tau = \frac{1}{\tau} \int_0^\tau dt\, \mathscr{L}\left(\rho_t, \rho_t^{eq}\right). \tag{3.48}$$

Due to the lengthy formula for the fidelity (3.47) it is not feasible to find a closed expression for $\langle \mathscr{L} \rangle_\tau$. Thus, we continue with a graphical analysis of the linear approximation. In order to determine the range of validity of the linear regime we, first, calculate the mean energy $\langle H \rangle$ in the linear approximation, and, second, compare it with the exact result (B.18). To this end, we identify the perturbation Hamiltonian as,

$$\mathscr{V}_t = -\frac{M}{2}(\omega_0^2 - \omega_t^2)x^2, \tag{3.49}$$

3 Dynamical properties of nonequilibrium quantum systems

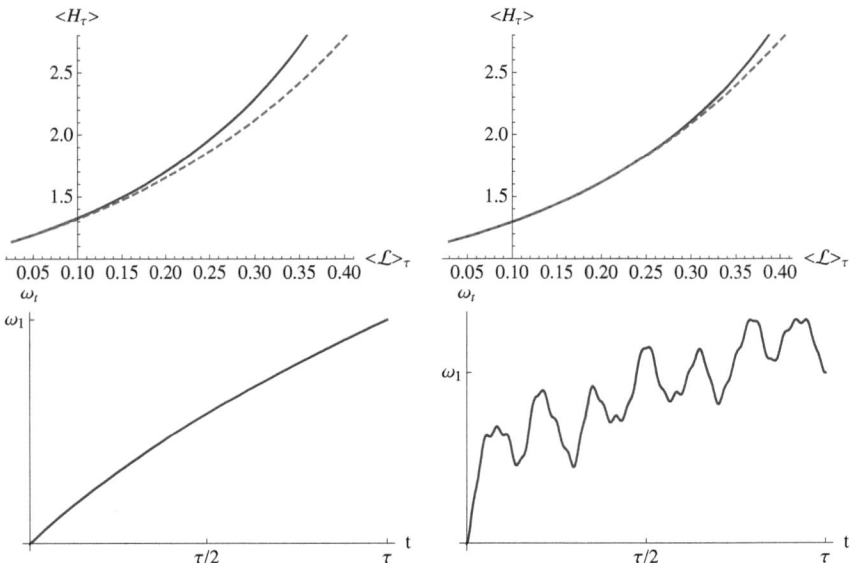

Figure 3.3: Upper plots: Exact mean energy (B.18) (dashed line) and linear approximation (3.50) (solid line) as a function of the time averaged Bures length (3.48) with $\beta\hbar\omega_0 = 1$, $M = 1$ and $\tau = 1$; Lower plots: corresponding parameterization of ω_t for arbitrary values of ω_1

and obtain by evaluating Eq. (3.37),

$$\langle H_\tau \rangle_{\text{linear}} = \frac{\hbar}{4} \frac{\omega_0^2 + \omega_1^2}{\omega_0^2} \coth\left(\frac{\beta}{2}\hbar\omega_0\right). \tag{3.50}$$

In Fig. 3.3 we plot the exact mean energy (B.18) and the linear approximation (3.50) as a function of the time averaged Bures length (3.48) for different parameterizations of ω_t. As expected the exact and the approximated mean energy are indistinguishable for small values of $\langle \mathscr{L} \rangle_\tau$. For larger values of $\langle \mathscr{L} \rangle_\tau$, i.e. for higher ω_1, the system is on average farther away from equilibrium, and, thus, the linear approximation fails. Concluding, the time averaged Bures length (3.48) is an appropriate measure for the validity of the linear regime.

Since the exact fidelity (3.47) is still a complicated function, we analyze in the following simplifying limits to deepen our insight.

3.2 Measuring the distance to equilibrium

Zero temperature limit

In the zero temperature limit, $\hbar\beta \gg 1$, where initially merely the ground state is occupied, the fidelity function reduces to,

$$F\left(\rho_t, \rho_t^{eq}\right) \xrightarrow{\hbar\beta \gg 1} \sqrt{\frac{2}{1+Q_t^*}}, \qquad (3.51)$$

which can be derived directly, as well. In the zero temperature limit the harmonic oscillator is initially in a pure state $|0\rangle$. Hence, the instantaneous equilibrium density operator reads at all times,

$$\rho_t^{eq}\Big|_{T=0} = |0\rangle\langle 0|. \qquad (3.52)$$

If one state is pure, the fidelity reduces to Eq. (3.27) and we obtain,

$$F\left(\rho_t, \rho_t^{eq}\right)\Big|_{T=0} = \operatorname{tr}\left\{\rho_t \rho_t^{eq}\right\} = \langle 0|\rho_t|0\rangle = \langle 0|U_t^\dagger \rho_0 U_t|0\rangle$$
$$= |\langle 0|U_t|0\rangle|^2 = p_{0,0}^t. \qquad (3.53)$$

In the latter equation $p_{0,0}^t$ denotes the transition probability from initial to final ground state, i.e. their overlap. It is given by [AL10],

$$p_{0,0}^\tau = \sqrt{\frac{2}{1+Q_t^*}}. \qquad (3.54)$$

Thus, we recover the zero temperature limit of the fidelity function (3.49), which was originally defined as the overlap of two wave functions (3.26).

Classical limit

On the other hand, in the limit of very high temperatures, i.e. in the classical limit, $\hbar\beta \ll 1$, we obtain,

$$F\left(\rho_t, \rho_t^{eq}\right)\Big|_{\hbar\beta \ll 1} = \frac{4\omega_0 \omega_t}{\omega_0^2 + 2Q_t^* \omega_0 \omega_t + \omega_t^2}. \qquad (3.55)$$

It is remarkable that in the classical limit the fidelity vanishes for large values of Q_t^* like $F_t \propto 1/Q_t^*$, whereas in the zero temperature limit $F_t \propto 1/\sqrt{Q_t^*}$.

3.3 Minimal quantum evolution time

In the last section we discussed an illustrative application of the Bures length. Now, we turn to a more fundamental implication of the Bures angle (3.30) for general, isolated quantum systems. Among the most interesting peculiarities of quantum mechanics are the Heisenberg uncertainty relations. They express the probabilistic nature of quantum systems, which distinguishes them from the classical world. Here, let us consider the Heisenberg uncertainty relation for energy and time,

$$\Delta E \, \Delta t \geq \hbar. \tag{3.56}$$

Usually, the latter equation is interpreted as: there is a minimal time it takes for a quantum system to evolve between two orthogonal states, which is always larger than the inverse of its initial energy spread ΔE. A more accurate expression for the *quantum speed limit*, τ_{QSL}, has been derived in [MT45, ML98, LT09] for time-independent Hamiltonians,

$$\tau_{\mathrm{QSL}} = \max\left\{\frac{\pi}{2}\frac{\hbar}{E}, \frac{\pi}{2}\frac{\hbar}{\Delta E}\right\}, \tag{3.57}$$

where E is the expectation value of the Hamiltonian minus the ground state energy, $E = \langle H \rangle - E_g$, and ΔE denotes the dispersion, $\Delta E^2 = \langle H \rangle^2 - \langle H^2 \rangle$. Only recently Giovannetti, Lloyd, and Maccone [LM03] treated the case of arbitrary angles within a numerical analysis, whereas Jones and Kok [JK10] proposed a systematic treatment based on a geometric approach for orthogonal states. However, all previous studies dealt merely with time-independent Hamiltonians. The following derivation of the quantum speed limit generalizes the geometric approach to arbitrary angles and explicitly time-dependent Hamiltonians. As argued earlier (cf. subsection 3.1.2) we can restrict ourselves without loss of generality to the case of pure states.

3.3.1 Mandelstam-Tamm type bound

Our derivation starts with the quantum Fisher information, which is an information theoretic quantity related to the Bures length (cf. appendix A.2). For two arbitrary density operators, ρ and ρ', and a parameterization ρ_t, which starts at ρ at $t = 0$

3.3 Minimal quantum evolution time

and ends at ρ' at $t = \tau$, the gradient of the Fisher information, $\mathscr{I}_t$, is given by,

$$d_t \mathscr{I}_t = (d_t \mathscr{L})^2 . \tag{3.58}$$

With the latter definition the distinction of two densities, ρ and ρ', is reduced to the estimation of t. The information about t obtained by a particular measurement is given by the Fisher information, $\mathscr{I}_t$. Moreover, it has been shown by Braunstein and Caves [BC94] that the Bures angle between two infinitesimally close density operators, $\rho' = \rho + d\rho$, i.e. the Bures metric (cf. appendix A.3) can be written as,

$$d\mathscr{L}^2 = \text{tr}\left\{d\rho\, \mathscr{R}_\rho^{-1}(d\rho)\right\} . \tag{3.59}$$

The superoperator $\mathscr{R}_\rho^{-1}$ reads in terms of the eigenvalues p_i of ρ, $\rho = \sum_i p_i |i\rangle\langle i|$,

$$\mathscr{R}_\rho^{-1}(O) = \sum_{j,k} \frac{\langle j|O|k\rangle}{p_j + p_k} |j\rangle |k\rangle, \tag{3.60}$$

for an arbitrary operator O. On the other hand, the von Neumann equation can be rewritten as,

$$i\hbar\, d_t \rho = [H, \rho] = [H - \langle H\rangle, \rho] = [\Delta H, \rho] , \tag{3.61}$$

where $\langle H \rangle$ is a real number, and can, hence, be included in the commutator. Finally, combining Eqs. (3.58)-(3.61) we conclude,

$$\begin{aligned}
d_t \mathscr{I}_t &= \text{tr}\left\{d_t \rho\, \mathscr{R}_\rho^{-1}(d_t \rho)\right\} = \frac{1}{\hbar^2} \sum_{j,k} \frac{(p_j - p_k)^2}{p_j + p_k} |\Delta H_{j,k}|^2 \\
&\leq \frac{1}{2\hbar^2} \sum_{j,k} (p_j + p_k) |\langle j|\Delta H_t|k\rangle|^2 = \frac{1}{\hbar^2} \left\langle \langle \Delta H \rangle^2 \right\rangle ,
\end{aligned} \tag{3.62}$$

where the latter estimation is implied by a triangle-type inequality. Equation (3.62) states that the amount of information about t in any measurement has an upper bound by the variance of its generator H. Now, the Mandelstam-Tamm type inequality follows almost immediately. With the definition of $\mathscr{I}_t$ (3.58) and taking the positive roots, we obtain,

$$d_t \mathscr{L} \leq \frac{1}{\hbar} \delta H , \tag{3.63}$$

3 Dynamical properties of nonequilibrium quantum systems

where $\delta H = \left| \sqrt{\langle \langle \Delta H \rangle^2 \rangle} \right|$. The latter differential inequality is solved by means of separating the variables and integration yields,

$$\int_0^{\mathscr{L}(\rho_0,\rho_\tau)} d\mathscr{L} \leq \frac{1}{\hbar} \int_0^\tau dt\, \delta H. \tag{3.64}$$

Thus, we finally obtain with the time averaged variance, $\Delta E_\tau = 1/\tau \int_0^\tau dt\, (\langle H_t^2 \rangle - \langle H_t \rangle^2)^{1/2}$, our first generalized quantum speed limit,

$$\tau \geq \frac{\hbar}{\Delta E_\tau} \mathscr{L}(\rho_0, \rho_\tau). \tag{3.65}$$

The latter result (3.65) generalizes the Mandelstam-Tamm bound [MT45] to arbitrary angles and time-dependent Hamiltonians. The derivation is merely based on the triangle-type inequality in Eq. (3.62) and the definition of the Bures angle in the space of density operators.

3.3.2 Margolus-Levitin type bound

Instead of using the bound on the Fisher information we can also use the finite expression of the angle $\mathscr{L}(\rho_0, \rho_\tau)$ between initial and final state to derive another differential inequality. To this end, we assume without loss of generality the system to be in a pure state at all times, $\rho_t = |\psi_t\rangle\langle\psi_t|$. Thus, we have,

$$\mathscr{L}(\psi_0, \psi_t) = \arccos\left(|\langle \psi_0 | \psi_t \rangle|\right). \tag{3.66}$$

Evaluating the absolute value of derivative of $\mathscr{L}(\psi_0, \psi_t)$ with respect to time, t, yields,

$$|d_t \mathscr{L}| = \frac{1}{\sqrt{1 - |\langle \psi_0 | \psi_t \rangle|^2}} |d_t|\langle \psi_0 | \psi_t\rangle|| = \frac{1}{\sin(\mathscr{L})} |d_t|\langle \psi_0 | \psi_t\rangle||. \tag{3.67}$$

Note, that $\sin(\mathscr{L}) > 0$, since $\mathscr{L} \in [0, \pi/2]$, and that in general $d_t \mathscr{L}$ is positive for short times, but may have either sign at different instants during the time span τ. In order to simplify the latter expression, we will prove that

$$|d_t|\langle \psi_0 | \psi_t\rangle|| \leq |d_t \langle \psi_0 | \psi_t\rangle|. \tag{3.68}$$

3.3 Minimal quantum evolution time

To this end, we expand the derivative on the left hand side of Eq. (3.68) as,

$$d_t |\langle \psi_0 | \psi_t \rangle| = d_t \sqrt{\langle \psi_0 | \psi_t \rangle \langle \psi_t | \psi_0 \rangle}. \tag{3.69}$$

With the help of the Schrödinger equation (3.1) the latter expression in Eq. (3.69) can be evaluated and we obtain,

$$|d_t |\langle \psi_0 | \psi_t \rangle|| = \left| \frac{\langle \psi_0 | \psi_t \rangle \langle \psi_t | H_t | \psi_0 \rangle - \langle \psi_0 | H_t | \psi_t \rangle \langle \psi_t | \psi_0 \rangle}{2i\hbar |\langle \psi_t | \psi_0 \rangle|} \right| \tag{3.70a}$$

$$= \left| \frac{\mathrm{Im}\left(\langle \psi_t | H_t | \psi_0 \rangle \langle \psi_0 | \psi_t \rangle \right)}{\hbar |\langle \psi_t | \psi_0 \rangle|} \right| \tag{3.70b}$$

$$\leq \frac{|\langle \psi_t | H_t | \psi_0 \rangle \langle \psi_0 | \psi_t \rangle|}{\hbar |\langle \psi_t | \psi_0 \rangle|}. \tag{3.70c}$$

On the other hand, $|d_t \langle \psi_0 | \psi_t \rangle|$ can be bounded from below by noting that,

$$|d_t \langle \psi_0 | \psi_t \rangle| = \frac{1}{\hbar} |\langle \psi_0 | H_t | \psi_t \rangle| = \frac{|\langle \psi_0 | H_t | \psi_t \rangle| |\langle \psi_t | \psi_0 \rangle|}{\hbar |\langle \psi_t | \psi_0 \rangle|}$$

$$\geq \frac{|\langle \psi_0 | H_t | \psi_t \rangle \langle \psi_t | \psi_0 \rangle|}{\hbar |\langle \psi_t | \psi_0 \rangle|}, \tag{3.71}$$

where we again used the Schrödinger equation (3.1). Furthermore, the inequality sign in Eq. (3.71) is implied by the Cauchy-Schwarz inequality. Finally, comparing Eqs. (3.70) and (3.71) leads to the desired estimation in Eq. (3.68). Continuing the derivation of the Margolus-Levitin type inequality, we combine Eqs. (3.68) and (3.71),

$$\sin(\mathscr{L}) |d_t \mathscr{L}| \leq |d_t \langle \psi_0 | \psi_t \rangle| = \frac{1}{\hbar} |\langle \psi_0 | H_t | \psi_t \rangle|. \tag{3.72}$$

The latter expression can be further estimated in absolute values, since the unitary time evolution (3.1) only yields a phase factor,

$$|\langle \psi_0 | H_t | \psi_t \rangle| = \left| \langle \psi_t | \mathscr{T}_< \exp\left(\frac{i}{\hbar} \int_0^\tau dt\, H_t \right) | \psi_t \rangle \right|$$

$$= \left| \sum_n |\langle \psi_t | n \rangle|^2 E_t^n \exp\left(\frac{i}{\hbar} \int_0^\tau dt\, E_t^n \right) \right| \tag{3.73}$$

$$\leq \left| \sum_n |\langle \psi_t | n \rangle|^2 E_t^n \right| = |\langle \psi_t | H_t | \psi_t \rangle|,$$

55

3 Dynamical properties of nonequilibrium quantum systems

where $\mathcal{T}_>$ denotes the time ordering operator and $\{|n\rangle\}$ is the set of instantaneous energy eigenstates, $H_t|n\rangle = E_t^n|n\rangle$. A final integration, with $|\int dx\, f(x)| \leq \int dx |f(x)|$ and $\sin(x) \geq 0$ for all $x \in [0, \pi/2]$, then results in

$$\int_0^{\mathscr{L}(\psi_0, \psi_\tau)} d\mathscr{L} \sin(L) \leq \frac{1}{\hbar} \int_0^\tau dt \, \langle \psi_t | H_t | \psi_t \rangle. \qquad (3.74)$$

Note, that the average energy is assumed to be positive at all times since the ground state energy is set to zero, $E_g = 0$. We have, therefore, obtained a generalized Margolus-Levitin uncertainty relation valid for arbitrary angles between initial and final pure states and arbitrary time-dependent driving,

$$\tau \geq \frac{\hbar}{E_\tau} |\cos(\mathscr{L}(\psi_0, \psi_\tau)) - 1| \geq \frac{4\hbar}{\pi^2 E_\tau} \mathscr{L}^2(\psi_0, \psi_\tau), \qquad (3.75)$$

where E_τ is the time averaged mean energy, $E_\tau = 1/\tau \int_0^\tau dt \, \langle \psi_t | H_t | \psi_t \rangle$. For the latter estimation in Eq. (3.75) we made use of the trigonometric inequality for the cosine function, $|\cos(x) - 1| \geq 4/\pi^2 x^2$ for all $x \in [0, \pi/2]$. The above derivation can be extended to arbitrary mixed states by interpreting pure states as purifications of mixed states in an enlarged Hilbert space [Joz94]. Specifically, by choosing the purification that maximizes the fidelity between two density operators, the Margolus-Levitin inequality also applies to the purifications [JK10]. Hence, we generally have,

$$\tau \geq \frac{4\hbar}{\pi^2 E_\tau} \mathscr{L}^2(\rho_0, \rho_\tau), \qquad (3.76)$$

with the mean energy $E_\tau = (1/\tau) \int_0^\tau dt \, \text{tr}\{\rho_t H_t\}$. It is worth mentioning, that Eq. (3.76) generalizes qualitatively the Margolus-Levitin-type bound put forward in [LM03] for time-independent systems. However, the numerical prefactor, $4/\pi^2$, is smaller than $2/\pi$ as in [LM03], since in contrast to [LM03] we have not numerically sharped the inequalities yielding the final estimation. Finally, we derive another version of a Margolus-Levitin-type bound by directly considering the

3.3 Minimal quantum evolution time

overlap of initial and final state. Hence, we have

$$
\begin{aligned}
|\langle \psi_0 | \psi_\tau \rangle| &= \left| \langle \psi_0 | \mathcal{T}_> \exp\left(-\frac{i}{\hbar} \int_0^\tau dt\, H_t\right) | \psi_0 \rangle \right| \\
&= \left| \sum_n |\langle \psi_0 | n \rangle|^2 \exp\left(-\frac{i}{\hbar} \int_0^\tau dt\, E_t^n\right) \right| \\
&\geq \left| \sum_n |\langle \psi_0 | n \rangle|^2 \cos\left(\frac{1}{\hbar} \int_0^\tau dt\, E_t^n\right) \right|,
\end{aligned}
\qquad (3.77)
$$

where the $\{|n\rangle\}$ is again the set of instantaneous energy eigenstates, $H_t |n\rangle = E_t^n |n\rangle$. The inequality sign in Eq. (3.77) is implied by estimating the absolute value from below by the real part only. Next, we express the cosine-function with its series representation, $\cos(x) = \sum_k (-1)^k / (2k)! \, x^{2k}$. By, moreover, using that $|\langle \psi_0 | n \rangle| \leq 1$ we conclude,

$$
\begin{aligned}
|\langle \psi_0 | \psi_\tau \rangle| &\geq \left| \sum_n \sum_k \frac{(-1)^k}{(2k)!} |\langle \psi_0 | n \rangle|^2 \left(\frac{1}{\hbar} \int_0^\tau dt\, E_t^n\right)^{2k} \right| \\
&\geq \left| \sum_n \sum_k \frac{(-1)^k}{(2k)!} |\langle \psi_0 | n \rangle|^{2k} \left(\frac{1}{\hbar} \int_0^\tau dt\, E_t^n\right)^{2k} \right| \\
&= \left| \sum_k \frac{(-1)^k}{(2k)!} \left(\frac{1}{\hbar} \int_0^\tau dt\, \langle \psi_0 | H_t | \psi_0 \rangle \right)^{2k} \right| \\
&= \left| \cos\left(\frac{1}{\hbar} \int_0^\tau dt\, \langle \psi_0 | H_t | \psi_0 \rangle\right) \right|.
\end{aligned}
\qquad (3.78)
$$

3 Dynamical properties of nonequilibrium quantum systems

Further, since the arccos (x) is a monotonically decreasing function for all $x \in [0, 1]$ we obtain

$$\arccos\left(|\langle\psi_0|\psi_\tau\rangle|\right) \leq \arccos\left(\left|\cos\left(\frac{1}{\hbar}\int_0^\tau dt\,\langle\psi_0|H_t|\psi_0\rangle\right)\right|\right)$$

$$\leq \left|\frac{1}{\hbar}\int_0^\tau dt\,\langle\psi_0|H_t|\psi_0\rangle\right| \leq \frac{1}{\hbar}\int_0^\tau dt\,|\langle\psi_0|H_t|\psi_0\rangle| \quad (3.79)$$

$$\leq \frac{1}{\hbar}\int_0^\tau dt\,\langle\psi_t|H_t|\psi_t\rangle,$$

where we used in the last inequality an argument analogous to Eq. (3.73). Hence, the time of quantum evolution from initial to final state can be estimated from below as

$$\tau \geq \frac{\hbar}{\overline{E_\tau}}\mathscr{L}(\psi_0,\psi_\tau). \quad (3.80)$$

As before the generalization to mixed quantum states is provided with the help of the appropriate partial traces.

3.3.3 Quantum speed limit

Collecting the above derived bounds on the quantum evolution time (3.66) and (3.80) we obtain the quantum speed limit τ_{QSL} as,

$$\tau_{\mathrm{QSL}} = \max\left\{\frac{\hbar\mathscr{L}(\rho_\tau,\rho_0)}{\overline{E_\tau}}, \frac{\hbar\mathscr{L}(\rho_\tau,\rho_0)}{\overline{\Delta E_\tau}}\right\}. \quad (3.81)$$

The minimum time is determined by the time averaged mean and variance of the energy and not by their initial values, as earlier estimated by numerical methods in [LM03],

$$\tau_{\min} \simeq \max\left\{\frac{2\hbar\mathscr{L}^2(\rho_\tau,\rho_0)}{\pi E_0}, \frac{\hbar\mathscr{L}(\rho_\tau,\rho_0)}{\Delta E_0}\right\}, \quad (3.82)$$

where $E_0 = \langle H_0\rangle$ and $\Delta E_0 = (\langle H_0^2\rangle - \langle H\rangle^2)^{1/2}$. The latter bound in Eq. (3.81) is merely valid for time-independent systems or quasistatic processes. As a consequence, the quantum speed limit time for driven systems can be smaller than

3.3 Minimal quantum evolution time

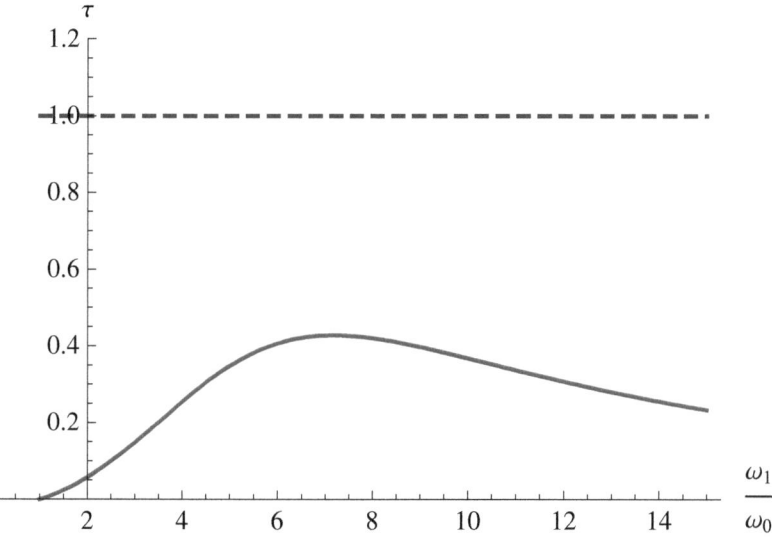

Figure 3.4: Quantum speed limit time τ_{QSL}, Eq. (3.81), (solid line) and actual driving time τ (dashed line) for the linearly parameterized quantum harmonic oscillator (3.40) and (3.83) with $\hbar = \tau = 1$, $1/\beta = 0$ and $\omega_0 = 0.1$.

for undriven systems, when $E_\tau > E_0$ ($\Delta E_\tau > \Delta E_0$). This is, for instance, the case at zero temperature: According to Eq. (3.82), a quantum system never leaves an initial (non-degenerate) pure state (infinite $\tau_{\min}$) in the absence of driving, while Eq. (3.81) predicts a finite τ_{QSL} for a driven Hamiltonian. Figure 3.4 shows that for the time-dependent, harmonic oscillator (3.40) at zero temperature, the actual driving time τ can approach the absolute minimum evolution time τ_{QSL} within a factor of two, for a simple linear change of its angular frequency,

$$\omega_t^2 = \omega_0^2 + (\omega_1^2 - \omega_0^2)t/\tau. \tag{3.83}$$

Hence, the quantum speed limit is attainable by properly choosing the protocol parameterizing the Hamiltonian. Moreover, it is remarkable that the rigorously derived quantum speed limit, τ_{QSL}, (3.81) does not exactly reduce for time-independent systems to the numerically estimated bound, $\tau_{\min}$, (3.82). The latter

3 Dynamical properties of nonequilibrium quantum systems

observation is not a contradiction, since the quantum speed limit is derived with the help of estimating inequalities, and, hence, with some degree of freedom.

The above derivation of the quantum speed limit (3.81) is valid for general driving and arbitrary angles. Therefore, earlier, heuristic assumptions [GS09, GC10] about the correct form of quantum speed limit time, τ_{QSL}, for time-dependent Hamiltonians are clarified.

Finally, in the classical limit, $\hbar \ll 1$, the quantum speed limit vanishes as the Heisenberg uncertainty relation (3.56) is a mere quantum effect.

3.4 Summary

The present chapter discussed dynamical properties of time-dependent quantum systems by means of a geometric approach. To this end, we introduced a natural distance on the space of density operators. The Bures length is, on the one hand, the generalization of the classical distinguishability distance to mixed quantum states, and, on the other hand, the generalized angle between density operators. Furthermore, we proposed the time averaged Bures length as an appropriate measure for the validity of the linear regime. A quantum system can be described by means of linear response theory, if it stays close to an equilibrium state at all times, i.e. if the time averaged Bures length is small. We confirmed this conclusion with the help of an analytically solvable system, namely the parameterized harmonic oscillator. Finally, we derived a fundamental lower bound on the time of quantum evolution. The quantum speed limit is the minimal time a quantum system needs to evolve between two distinguishable states. This minimal time is governed by the time averaged mean and variance of the system energy and the Bures length, i.e. the angle between initial and final state.

4 Unitary quantum processes in thermally isolated systems

Chapter 3 dealt with the dynamical properties of time-dependent quantum systems. In particular, the meaning of equilibrium and nonequilibrium states and their distinguishability were clarified. In the present chapter we come back to thermodynamics. As before, we assume the quantum system under consideration to be isolated, and, thus, describable by means of unitary dynamics (3.1). We will start with a preliminary discussion of the notion of work and heat in quantum systems before we turn to a generalized and sharpened Clausius inequality. We will see that the geometric approach, which was discussed in detail in the last chapter, turns out to be an appropriate method to explain the thermodynamics of closed quantum systems.

4.1 Thermodynamics: Work and heat in quantum mechanics

Let us start with two basic quantities in thermodynamics, work, W, and heat, Q. Since W and Q are path or trajectory dependent their infinitesimal increments, δW and δQ, are not given by total differentials. This fact is a first hint that a careful treatment is necessary in quantum mechanics. In quantum systems the meaning of single trajectories is lost, and one rather has to deal with probability distributions. How to define work and heat quantum mechanically, and the according corollaries are discussed in this section. Moreover, we assume as before in the context of classical Hamiltonian dynamics (in subsection 2.3.1) the system to be initially thermalized, but otherwise isolated during the process in a time interval, $0 \leq t \leq \tau$.

4 Unitary quantum processes in thermally isolated systems

4.1.1 Work is not an observable

The discussion of the probability density of quantum work was initiated by Kurchan [Kur00] and Tasaki [Tas00]. The conclusions presented here follow Talkner, Lutz, and Hänggi [LH07]. We consider an explicitly time-dependent quantum system with Hamiltonian H_t (cf. Eq. (3.1)). Thus, thermodynamic work is performed during the variation of H_t with time. Microscopically, quantum work for a single transition between distinct eigenstates of the system is given by the difference of the initial and final energy eigenvalue, E_n^0 and E_m^τ. For an initial density, $\rho_0 = \exp(-\beta H_0)/Z_0$ with eigenvalues p_n^0, we have to average over the thermal occupation probabilities. Furthermore, we have to account for the induced transitions by the unitary evolution with probabilities $p_{m,n}^\tau$. Hence, the probability density of the total work done on the quantum system during time τ is given by,

$$\mathscr{P}(W) = \sum_{m,n} \delta\left(W - (E_m^\tau - E_n^0)\right) p_{m,n}^\tau p_n^0. \qquad (4.1)$$

Note that Eq. (4.1) expresses the work being a random quantity due to the presence of both thermal and quantum uncertainties, encoded in p_n^0 and $p_{m,n}^\tau$, respectively. The transition probabilities are given by the overlap between an instantaneous eigenstate, $H_\tau |m_\tau\rangle = E_m^\tau |m_\tau\rangle$, and a time evolved initial one, $H_0 |n_\tau\rangle = E_n^0 |n_0\rangle$,

$$p_{m,n}^\tau = |\langle m_\tau | U_\tau | n_0 \rangle|^2, \qquad (4.2)$$

where the time evolution operator is, as usual, given with the time ordering operator, $\mathscr{T}_>$, by,

$$U_\tau = \mathscr{T}_> \exp\left(\frac{1}{i\hbar} \int_0^\tau dt\, H_t\right). \qquad (4.3)$$

Equivalently to the probability distribution, $\mathscr{P}(W)$ in Eq. (4.1), the statistical properties of W are encoded in the characteristic function, $\mathscr{G}(\mu)$, which is defined as the Fourier transform of the probability density,

$$\mathscr{G}(\mu) = \int dW\, \exp(i\mu W)\, \mathscr{P}(W). \qquad (4.4)$$

4.1 Thermodynamics: Work and heat in quantum mechanics

Substituting the probability density, $\mathscr{P}(W)$ in Eq. (4.1), with Eq. (4.2) into Eq. (4.4), the characteristic function, $\mathscr{G}(\mu)$, can be evaluated to read,

$$\begin{aligned}\mathscr{G}(\mu) &= \sum_{m,n} \exp\left(i\mu\left(E_m^\tau - E_n^0\right)\right) \langle m_\tau | U_\tau | n_0\rangle \langle n_0 | U_\tau^\dagger | m_\tau\rangle \frac{1}{Z_0} \exp(-\beta E_n^0) \\ &= \sum_{m,n} \langle m_\tau | U_\tau \exp(-i\mu H_0) \rho_0 | n_0\rangle \langle n_0 | U_\tau^\dagger \exp(i\mu H_\tau) | m_\tau\rangle .\end{aligned} \quad (4.5)$$

By further making use of the completeness relation of the energy eigenstates, the characteristic function, $\mathscr{G}(\mu)$, is written as an average over the initial density, ρ_0,

$$\begin{aligned}\mathscr{G}(\mu) &= \operatorname{tr}\left\{U_\tau^\dagger \exp(i\mu H_\tau) U_\tau \exp(-i\mu H_0) \rho_0\right\} \\ &= \langle \exp(i\mu H_\tau) \exp(-i\mu H_0) \rangle_{\rho_0} .\end{aligned} \quad (4.6)$$

We observe that the characteristic function, $\mathscr{G}(\mu)$ in Eq. (4.6), has the form of a time-ordered two point correlation function. Therefore, we conclude that there is no Hermitian operator defining the work in quantum systems. This is in agreement with the classical, thermodynamic notion of work being not a state function. In order to determine the work, a two-time measurement is necessary. Since the Hamiltonian at different times does not have to commute with itself, the characteristic function (4.6) cannot be further simplified without an additional time-ordering operator [LH07]. The latter point becomes more obvious by the observation that the characteristic function equals the exponentiated work (cf. Eq. (4.4)) for the particular choice, $\mu = i\beta$. By further evaluating Eq. (4.6) we obtain the quantum version of the Jarzynski equality (2.36),

$$\langle \exp(-\beta W) \rangle = \langle \exp(-\beta H_\tau) \exp(\beta H_0) \rangle_{\rho_0} = \frac{Z_\tau}{Z_0} = \exp(-\beta \Delta F) . \quad (4.7)$$

The main conclusion is that the Jarzynski equality (4.7) remains valid for isolated quantum systems. However, the nonequilibrium work is not an observable and, thus, one has to concentrate on quantum statistical properties of thermodynamic quantities. To this end, the present chapter provides a detailed analysis of implications and applications of the work density (4.1).

4 Unitary quantum processes in thermally isolated systems

4.1.2 Fluctuation theorem for heat exchange

In the last subsection it was shown that there is no Hermitian operator describing the work performed during a quantum process. In the latter analysis we restricted ourselves to the simple case of unitary dynamics, and, thus, isolated systems. This means explicitly that no heat was exchanged with any environment. Now, we turn our attention to situations in which merely heat is exchanged, but no work is performed. The present analysis follows Jarzynski and Wójcik [JW05] in our notation. Let us consider a quantum system consisting of two subsystems,

$$H^{\text{tot}} = H^A \otimes \mathbb{1}^B + \mathbb{1}^A \otimes H^B + h_\gamma, \tag{4.8}$$

where h_γ denotes a very weak interaction between A and B. Moreover, we consider the total Hamiltonian, H^{tot}, to be time-independent in order to exclude work performing processes. Now, we assume that the total system and both subsystems are time-reversal invariant. In quantum mechanics the invariance of time reversal of a system is expressed by the condition,

$$[\Theta, H] = 0, \tag{4.9}$$

where Θ is the quantum time-reversal operator and H the Hamiltonian of the system under consideration. The operator Θ reverses linear and angular momentum while keeping position unchanged. Now, we assume the subsystems A and B having equilibrated with inverse temperatures β_A and β_B before the experiment and are, thus, described by thermal densities, $\rho^i = \exp\left(-\beta_i H^i\right)/Z_i$ with $i = A, B$. At time $t = 0^-$ we separate the systems from the reservoirs having induced the equilibration. By measuring the energies each subsystem i is projected onto a pure state $|n_i\rangle$ with probability $p_{n^i}^i = \exp\left(-\beta_i E_{n^i}^i\right)/Z_i$ and the total system is described by the product state $|n^A\rangle \otimes |n^B\rangle$. Turning on the interaction term, h_γ, at $t = 0$ we allow the system to evolve until $t = \tau$. The final state, $|\psi_\tau\rangle$, of the total system is determined by the unitary evolution under Schrödinger's equation for the total system (3.1). Now, we turn off the interaction term, and measure once again the energies of each subsystem separately. Therefore, the state $|\psi_\tau\rangle$ is projected onto a final product state $|m^A\rangle \otimes |m^B\rangle$. It is worth emphasizing that the time τ may be chosen randomly. Usually, the total system is in an arbitrary nonequilibrium state at time τ. Since we assumed weak coupling between the subsystems, we expect the energy of the total system to be almost preserved,

$$E_n^A + E_n^B \simeq E_m^A + E_m^B. \tag{4.10}$$

4.1 Thermodynamics: Work and heat in quantum mechanics

Hence, the energy changes of the two subsystems are approximately equal and we identify the heat exchange $Q_{n \to m}$ as,

$$Q_{n \to m} = E_m^B - E_n^B \simeq E_n^A - E_m^A. \tag{4.11}$$

Analogously to the work density (4.1) we, now, formulate the probability density of the heat exchange, Q, to read,

$$\mathscr{P}(Q) = \sum_{n,m} \delta\left(Q - Q_{n \to m}\right) p_{m,n}^\tau p_{n^A}^A p_{n^B}^B, \tag{4.12}$$

where $|n\rangle = |n^A\rangle \otimes |n^B\rangle$ and $|m\rangle = |m^A\rangle \otimes |m^B\rangle$. As before, $p_{m,n}^\tau$ denotes the transition probability between a time evolved initial state $U_\tau |n\rangle$ and the final energy eigenstate $|m\rangle$ (cf. Eq. (4.2)). Here, U_τ denotes the unitary time evolution operator of the total system. In order to derive the fluctuation theorem for the heat transfer, Q, we, now, consider the total probability for a forward transition from $|n\rangle$ to $|m\rangle$,

$$P_\tau(|n\rangle \to |m\rangle) = p_{m,n}^\tau p_{n^A}^A p_{n^B}^B = |\langle m_\tau | U_\tau | n_0 \rangle|^2 \frac{\exp\left(-\beta_A E_{n^A}^A\right) \exp\left(-\beta_B E_{n^B}^B\right)}{Z_A Z_B}, \tag{4.13}$$

and its time reversed counterpart (cf. Eqs. (2.50) and (2.51)),

$$P_\tau(\Theta | n \rangle \to \Theta | m \rangle) = \left|\langle m_\tau | \Theta^\dagger U_\tau \Theta | n_0 \rangle\right|^2 \frac{\exp\left(-\beta_A E_{m^A}^A\right) \exp\left(-\beta_B E_{m^B}^B\right)}{Z_A Z_B}. \tag{4.14}$$

The time-reversal operator Θ, however, is anti-unitary, and $U_\tau \Theta = \Theta U_{-\tau}$, which follows from Eq. (4.9) and the anti-linearity [JW05]. Hence, we conclude for the transition probabilities,

$$\langle m_\tau | \Theta^\dagger U_\tau \Theta | n_0 \rangle = \langle m_\tau | \Theta^\dagger \Theta U_{-\tau} | n_0 \rangle = \langle m_\tau | U_{-\tau} | n_0 \rangle, \tag{4.15}$$

and for the ratio of forward and reversed transition,

$$\frac{P_\tau(|n\rangle \to |m\rangle)}{P_\tau(\Theta|n\rangle \to \Theta|m\rangle)} \simeq \exp\left(\Delta \beta \, Q_{n \to m}\right). \tag{4.16}$$

In the latter equation we introduced the temperature difference, $\Delta \beta = \beta_B - \beta_A$. Due to the unitary evolution of the total system each eigenstate has a corresponding time-reversed twin. Thus, the net probability of the heat transfer Q in time τ

4 Unitary quantum processes in thermally isolated systems

can be written as,

$$\mathscr{P}(Q) = \sum_{n,m} \delta(Q - Q_{n \to m}) P_\tau(|n\rangle \to |m\rangle)$$
$$= \exp(\Delta\beta Q) \sum_{\Theta n, \Theta m} \delta(Q + Q_{\Theta n \to \Theta m}) P_\tau(\Theta|n\rangle \to \Theta|m\rangle) \quad (4.17)$$
$$= \exp(\Delta\beta Q) \mathscr{P}(-Q).$$

In Eq. (4.17) we obtained a fluctuation theorem for the heat transfer Q between to weakly coupled systems. The weak coupling limit is crucial for the identification of the heat, where we assumed that no energy is lost in the interaction, h_γ. Furthermore, the weak coupling limit ensures that the final state $|\psi_\tau\rangle$ can be projected onto a product state of the two subsystems without loss of information.

Illustrative example - two coupled harmonic oscillators

Next, in order to illustrate $\mathscr{P}(Q)$ in Eq. (4.12), we consider an analytically solvable example, namely two weakly coupled, isotropic harmonic oscillators. The total system Hamiltonian reads,

$$H^{\text{tot}} = H^A \otimes \mathbb{1}^B + \mathbb{1}^A \otimes H^B + \gamma x^A \otimes x^B, \quad (4.18)$$

where the Hamiltonian of each subsystem is quadratic, $H^i = p_i^2/2M + M/2\,\omega^2 x_i^2$, $i = A, B$. For small coupling coefficients, $\gamma \ll M\omega^2$, the transition probabilities, $p_{m,n}^\tau$, can be calculated by means of time-independent perturbation theory [DL77a]. It can be shown that second order perturbation theory is sufficient to obtain numerically exact results for the probability density $\mathscr{P}(Q)$ [LD10]. Since the perturbational distribution, $\mathscr{P}(Q)$, is given by a lengthy formula, we, here, merely present illustrative plots based on a numerical analysis. Later on (cf. subsection 4.5.3), we will discuss the perturbational approach to the work distribution (4.1), as well. The heat distribution (4.12) is given by a sum of δ-peaks. Hence, the cumulative distribution,

$$\mathscr{P}_{\text{int}}(Q) = \int_{-\infty}^{Q} dQ'\, \mathscr{P}(Q'), \quad (4.19)$$

is given by a sum of step functions. In Fig. 4.1 we plot $\mathscr{P}_{\text{int}}(Q)$ for different time spans τ. We choose initial conditions, where the subsystem A starts with

4.2 Generalized Clausius inequality

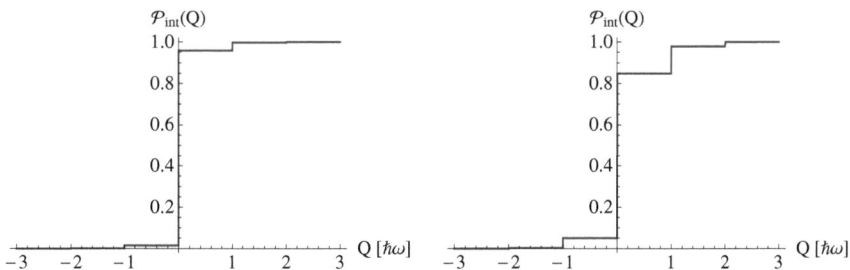

Figure 4.1: Cumulative heat distribution $\mathscr{P}_{\text{int}}(Q)$ (4.19) with $\gamma = \hbar\omega/20$, $\beta_A = 2$, $\beta_B = 1$, and $\tau = 10$ (left) and $\tau = 20$ (right).

smaller temperature than subsystem B, i.e. $\beta_A > \beta_B$. We observe that the mean heat flux is larger than zero, as B is in a *hotter* state than A. However, due to the small system size fluctuations are important and, thus, negative values for the heat transfer occur. That means explicitly that there is a nonzero probability that for single realizations heat flows from cold (subsystem A) to hot (subsystem B). Furthermore, the mean heat exchange, $\langle Q \rangle$, grows with time and the distributions broaden by increasing the process time, i.e. the longer the subsystems interact, the more account fluctuations.

4.2 Generalized Clausius inequality

In the last section we introduced the quantum mechanical notions of heat and work. We analyzed, why we cannot define an operator describing work or heat. Hence, we concentrated on the statistical properties. For nonequilibrium situations, however, the irreversible entropy production characterizes the dynamics. We saw earlier (cf. section 3.2) that the time averaged Bures length is appropriately measuring the distance of a nonequilibrium to an equilibrium state. Now, we return to a thermodynamic approach by generalizing the Clausius inequality (2.4) to quantum processes. Classical thermodynamics, however, merely states that the irreversible entropy production during an arbitrary process is always nonnegative [Pri47], $\Delta S_{\text{ir}} \geq 0$, with equality for quasistatic processes. The validity of the latter Clausius inequality is independent of the kind of operation on the system. For highly nonquasistatic transformations, however, the information gained

4 Unitary quantum processes in thermally isolated systems

about the size of ΔS_{ir} is not very sharp. Hence it would be desirable to generalize and sharpen the Clausius inequality for isolated quantum systems undergoing processes arbitrarily far from equilibrium (see also [DL10]). To this end, we, first, identify the irreversible entropy production, S_{ir}, and second, sharpen the Clausius inequality by deriving a lower bound in terms of the Bures length (3.30) between the final nonequilibrium and equilibrium states. Furthermore, we assume the system to be isolated during the process under consideration. Nevertheless, initially the quantum system starts in a thermal equilibrium state. The latter assumption is justified for systems which are ultra-weakly coupled to a thermal environment. Then the time scales of the interaction between system and bath are infinitely large compared to the time scales of the considered nonequilibrium processes. Hence, the dynamics of the system during such a process can be described by unitary dynamics, whereas for quasistatic driving the system remains in thermal equilibrium at all times.

4.2.1 Irreversible entropy production

We start by deriving an analytic expression for the irreversible entropy production from microscopic principles. Usually the entropy production, ΔS_{ir}, is identified with the mean irreversible part of the work [Cro98],

$$\beta \langle W_{\text{ir}} \rangle = \Delta S_{\text{ir}}, \qquad (4.20)$$

where $\langle W_{\text{ir}} \rangle = \langle W \rangle - \Delta F$. Here, ΔF is the equilibrium work performed during quasistatic processes. It is worth emphasizing that we compare a fast, nonequilibrium situation, described by W, with its thermodynamic equivalent ΔF. The equilibrium process, however, cannot be realized by unitary dynamics, since the system relaxes infinitely fast to equilibrium. As discussed earlier quantum mechanically work is not an observable (cf. subsection 4.1.1). It is rather given by a time-ordered correlation function and, hence, special interest lies on the probability distribution (4.1),

$$\mathscr{P}(W) = \sum_{m,n} \delta\left(W - \left(E_m^\tau - E_n^0\right)\right) p_{m,n}^\tau p_n^0. \qquad (4.21)$$

Here, $\left(E_m^\tau - E_n^0\right)$ is the microscopic work defined by the difference of initial and final energy eigenvalues. Starting with a canonical preparation one has to average over the given initial (thermal) density operator $\rho_0 = \exp(-\beta H_0)/Z_0$ with

4.2 Generalized Clausius inequality

eigenvalues p_n^0. Further, quantum mechanics introduces additional uncertainty by possible transitions with probabilities $p_{m,n}^\tau$ between initial and final states $|n_0\rangle$ and $|m_\tau\rangle$. The derivation of the expression for the irreversible entropy production starts with the averaged work,

$$\langle W \rangle = \sum_{m,n} \left(E_m^\tau - E_n^0 \right) p_{m,n}^\tau p_n^0. \tag{4.22}$$

We define the thermal density operator at time τ as, $\rho_\tau^{eq} = \exp(-\beta H_\tau)/Z_\tau$ with eigenvalues p_m^τ, which corresponds to the equilibrium state of the final configuration of the system. Thus, Eq. (4.22) can be rewritten,

$$\beta \langle W \rangle = \sum_n p_n^0 \ln p_n^0 - \sum_{m,n} p_n^0 p_{m,n}^\tau \ln p_m^\tau + \ln(Z_0/Z_\tau), \tag{4.23}$$

and we obtain with the definition of the free energy, $\Delta F = -1/\beta \ln(Z_\tau/Z_0)$, and an expression of the second law, $\langle W \rangle = \langle W_{\text{ir}} \rangle + \Delta F$, that the irreversible part of the work is given by,

$$\beta \langle W_{\text{ir}} \rangle = \sum_n p_n^0 \ln p_n^0 - \sum_{m,n} p_n^0 p_{m,n}^\tau \ln p_m^\tau. \tag{4.24}$$

On the other hand, the quantum Kullback-Leibler divergence [KL51, Ume62], the relative entropy, is defined as,

$$S\left(\rho_\tau || \rho_\tau^{eq}\right) = \text{tr}\left\{\rho_\tau \ln \rho_\tau - \rho_\tau \ln \rho_\tau^{eq}\right\}. \tag{4.25}$$

The relative entropy, $S\left(\rho_\tau || \rho_\tau^{eq}\right)$, is a non-commutative measure of the distinction between two density operators, ρ_τ and ρ_τ^{eq}. Moreover, $S\left(\rho_\tau || \rho_\tau^{eq}\right) \geq 0$ with equality only for identical densities (cf. appendix A.1). Equation (4.20) can be evaluated by applying the equilibrium density operators explicitly and we conclude by comparing Eqs. (4.23) and (4.25),

$$\beta \langle W \rangle = S\left(\rho_\tau || \rho_\tau^{eq}\right) + \beta \Delta F, \tag{4.26}$$

and especially

$$\Delta S_{\text{ir}} = \beta \langle W_{\text{ir}} \rangle = S\left(\rho_\tau || \rho_\tau^{eq}\right) \geq 0, \tag{4.27}$$

which constitutes our first quantum generalization of the second law of thermodynamics. Arbitrarily far from equilibrium the irreversible entropy production is

4 Unitary quantum processes in thermally isolated systems

given by the relative entropy measuring the distinguishability of the final nonequilibrium density, ρ_τ, and its equilibrium counter part, ρ_τ^{eq}. By further noting the invariance of $S\left(\rho_\tau || \rho_\tau^{eq}\right)$ under unitary transformations [Ume62] we rediscover in the classical limit the results of [PdB07] and [VJ09]. We note, however, that the relative entropy is not a true metric, as it is not symmetric and does not satisfy the triangle inequality; it, therefore, cannot be used as a proper quantum distance. Moreover, the exact result in Eq. (4.27) is not very suitable for practical use. For the evaluation of the relative entropy the full density operators are needed. Hence, we next derive a lower bound for the quantum entropy production, which we express in terms of the fidelity, one of the most commonly used and well-studied measures in quantum information theory [NC00].

4.2.2 Lower bound for the irreversible entropy

The relative entropy, $S(\rho_1||\rho_2)$, measures the distinguishability of two density operators, ρ_1 and ρ_2. However, $S(\rho_1||\rho_2)$ is neither symmetric nor fulfills a triangle inequality. In an earlier chapter (cf. subsection 3.1.2) we discussed the Bures length as a proper distance on the space of density operators. Hence, we would like to estimate the irreversible entropy production (4.27) from below with the help of the Bures length, $\mathscr{L}$. To this end we, first, analyze the Kullback-Leibler divergence, $D(p_1||p_2)$, i.e. the classical relative entropy of two probability measures, p_1 and p_2.

Lower bound in terms of the Bures distance

Let us start by introducing the Hellinger distance, h. Like Wootters' statistical distance (3.17) the Hellinger distance is a measure of the distinguishability of two probability distributions. It is, as well, a true distance, which fulfills the mathematical conditions, i.e symmetry, non-negativity and the triangle inequality. For two probability distributions, p_1 and p_2, h is defined as,

$$h^2(p_1, p_2) = \int dx \left(\sqrt{p_1(x)} - \sqrt{p_2(x)}\right)^2. \tag{4.28}$$

Definition (4.28) can be rewritten in terms of the continuous, classical fidelity function (3.18), $f(p_1, p_2) = \int dx \sqrt{p_1(x) p_2(x)}$, as,

$$h(p_1, p_2) = \sqrt{2 - 2f(p_1, p_2)}. \tag{4.29}$$

4.2 Generalized Clausius inequality

The corresponding Kullback-Leibler divergence between p_1 and p_2 can, now, be estimated from below. With the functional estimation, $\sqrt{y} - 1 \geq 1/2 \ln(y)$, we write,

$$\sqrt{\frac{p_2(x)}{p_1(x)}} - 1 \geq \frac{1}{2} \ln\left(\frac{p_2(x)}{p_1(x)}\right). \tag{4.30}$$

The latter equation has to be averaged and we obtain by taking expectation values with respect to p_1,

$$2\left(1 - \left\langle \sqrt{\frac{p_2(x)}{p_1(x)}} \right\rangle_{p_1}\right) \leq \left\langle \ln\left(\frac{p_1(x)}{p_2(x)}\right) \right\rangle_{p_1}. \tag{4.31}$$

Hence, we conclude that the Kullback-Leibler divergence of two probability distributions, p_1 and p_2, is bounded from below by the square of the Hellinger distance between p_1 and p_2,

$$D(p_1 \| p_2) \geq 2 - 2f(p_1, p_2) = h^2(p_1, p_2). \tag{4.32}$$

Next, we generalize the classical bound (4.32) to density operators involving mixed quantum states. The quantum generalization of the Kullback-Leibler divergence, $D(p_1 \| p_2)$, is given by the relative entropy, $S(\rho_1 \| \rho_2)$. Moreover, the quantum analog of the classical Hellinger distance (4.28) is given by the Bures distance, $\mathscr{D}$. Here, the Bures distance is a further element of the distance family implied by the Bures metric (cf. appendix A.3). Thus, $\mathscr{D}$ reads in terms of the quantum fidelity (3.19),

$$\mathscr{D}^2(\rho_1, \rho_2) = 2\left(1 - \sqrt{F(\rho_1, \rho_2)}\right). \tag{4.33}$$

As we discussed earlier (cf. subsection 3.1.2) the fidelity function for mixed quantum states is given by the overlap of pure states in an enlarged Hilbert space. Hence, the quantum version of the expectation values in Eq. (4.30) are obtained by taking the partial trace over the purifications of ρ_1 and ρ_2 and we obtain,

$$2\left(\sqrt{F(\rho_1, \rho_2)} - 1\right) \leq \langle \ln(\rho_1) - \ln(\rho_2) \rangle_{\rho_1}. \tag{4.34}$$

Concluding, the relative entropy and, therefore, the irreversible entropy production (4.27) is bounded from below by the Bures distance between the final nonequilibrium state, ρ_τ, and its equilibrium counter part, ρ_τ^{eq},

$$\Delta S_{\text{ir}} \geq \mathscr{D}^2\left(\rho_\tau, \rho_\tau^{\text{eq}}\right). \tag{4.35}$$

4 Unitary quantum processes in thermally isolated systems

The latter equation constitutes our first generalization of the Clausius inequality. In other words, inequality (4.35) quantifies in a precise way the intuitive notion that the irreversible entropy production is larger, when a system is driven farther away from equilibrium. However, the lower bound in terms of the Bures distance is merely useful for nonequilibrium processes which drive the system not to far away from equilibrium. The Bures distance is bounded from above, whereas the relative entropy grows exponentially. Moreover, the Bures distance, $\mathscr{D}$, lacks a clear physical interpretation in contrast to the Bures length, $\mathscr{L}$, being the generalized angle between mixed quantum states. Hence, we sharpen the latter inequality (4.35) in the following and express the lower bound of the irreversible entropy production in terms of $\mathscr{L}$ (3.30).

Lower bound in terms of the Bures length

Only recently Audenaert and Eisert established a mathematical, lower bound for the relative entropy [AE05]. For any unitarily invariant norm, $d(\rho_1, \rho_2)$, it can be shown that,

$$S(\rho_1||\rho_2) \geq s\left(\frac{d(\rho_1,\rho_2)}{d(e^{1,1},e^{2,2})}\right), \qquad (4.36)$$

where $e^{i,j} = |i\rangle\langle j|$ are the projection operators, i.e. the matrices with the i, j element equal to 1 and all other elements 0. The function $s(.)$ is obtained by an optimization procedure and can, eventually, be written as,

$$s(x) = \min_{x<r<1}\left\{(1-r+x)\log\left(1+\frac{x}{1-r}\right) + (r-x)\log\left(1-\frac{x}{r}\right)\right\}. \qquad (4.37)$$

Generally, the minimum on the right hand side of Eq. (4.37) has to be evaluated numerically. However, it can be expanded in series around $x=0$, which converges quickly. The leading orders of the series expansion read,

$$s(x) = 2x^2 + \frac{4}{9}x^4 + \frac{32}{135}x^6 + \frac{992}{5103}x^8 + \frac{6656}{32805}x^{10} + O(x^{12}). \qquad (4.38)$$

For a discussion of the convergence of the series we refer to the literature [AE05]. Note, however, that the latter expansion (4.38) is more precise than earlier published versions due to the evaluation of two higher orders.

Now, we can choose an arbitrary, unitarily invariant norm, d, to evaluate the sharp lower bound of the relative entropy (4.36). Thermodynamically relevant

4.2 Generalized Clausius inequality

distances should be physically motivated and to some degree unique [Rup95]. Since we saw earlier that the Bures length, $\mathscr{L}$ (3.30), is an appropriate measure for the distance between equilibrium and nonequilibrium states, we proceed with $s(\mathscr{L})$. Noting that $\mathscr{L}(e^{1,1}, e^{2,2}) = \pi/2$, since the two matrices are orthogonal ($F(e^{1,1}, e^{2,2}) = 0$), we obtain the generalized Clausius inequality,

$$\Delta S_{\text{ir}} \geq s\left(\frac{2}{\pi}\mathscr{L}\left(\rho_\tau, \rho_\tau^{\text{eq}}\right)\right) \geq \frac{8}{\pi^2}\mathscr{L}^2\left(\rho_\tau, \rho_\tau^{\text{eq}}\right), \qquad (4.39)$$

where the last term on the right hand side of Eq. (4.39) is the leading quadratic order. The latter Eq. (4.39) constitutes a sharp lower bound on the irreversible entropy production. Consequently, Eq. (4.39) is the properly generalized Clausius inequality for nonquasistatic quantum processes. The loss during an arbitrary process can be estimated from below with the length measuring the distance between the current state and the corresponding equilibrium one. From an experimental point of view the generalized Clausius inequality (4.39) estimates the quality of a process with the help of the fidelity. Moreover, we conclude that the minimal irreversible entropy production is governed by the generalized angle between the final nonequilibrium state and its corresponding equilibrium one. The higher orders in the series expansion are essential for a sharp lower bound, since for large excitations the angle converges to its maximal value $\pi/2$, whereas ΔS_{ir} grows continuously.

In Fig. 4.2 we compare the above, constructively derived lower bound in terms of the Bures distance, $\mathscr{D}$ in Eq. (4.35), with the leading order of the series expansion in terms of the Bures length, $\mathscr{L}$ in Eq. (4.39). We observe that both distance measures behave qualitatively similarly as both are implied by the same underlying metric (cf. appendix A.3). In the leading order, the Bures distance (4.35) sets a sharper lower bound on the irreversible entropy production. However, the Bures length, i.e. the angle between the final nonequilibrium state and its equilibrium counter part, has a clear physical interpretation, whereas the Bures distance is merely a related mathematical quantity. Finally, the constructively derived lower estimation (4.35) is recovered as leading order of the series expansion (4.38) in terms of the Bures distance.

For almost quasistatic transformations, where the system remains close to an equilibrium state at all times, the entropy production can be approximated by the

4 Unitary quantum processes in thermally isolated systems

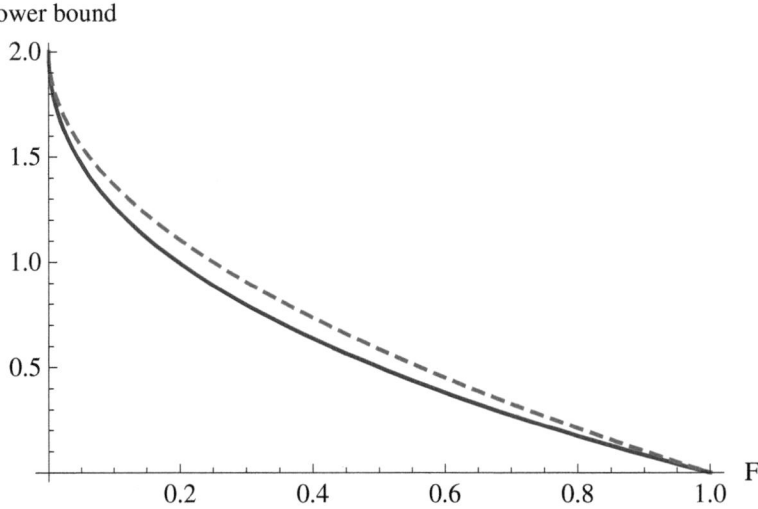

Figure 4.2: Lower bounds for the entropy production $8/\pi^2 \mathscr{L}^2\left(\rho_\tau, \rho_\tau^{eq}\right)$ (solid line) and $\mathscr{D}^2\left(\rho_\tau, \rho_\tau^{eq}\right)$ (dashed line) as a function of the fidelity $F\left(\rho_\tau, \rho_\tau^{eq}\right)$.

infinitesimal Bures length, $\mathscr{L}(\rho^{eq} + d\rho)$ [LP05],

$$4S(\rho^{eq} + d\rho \| \rho^{eq}) \simeq \mathscr{L}^2(\rho^{eq} + d\rho, \rho^{eq}). \tag{4.40}$$

The latter Eq. (4.40) constitutes with Eq. (4.27) the quantum generalization of Salamon's infinitesimal expression for the classical entropy production published almost thirty years ago [NB84, SB83, AA85]. Whereas in the classical case the generalized Clausius inequality easily follows from integrating the classical equivalent of Eq. (4.40) and the Cauchy-Schwarz inequality [Cro07], here, we have to deal with infinitesimal changes in the operator space. Concluding, our generalized Clausius inequality (4.39) implies in the classical limit extensions of previously proposed results. So far the entropy production was estimated [SB83, NB84, AA85, Cro07] from below by the square of the thermodynamic length. Earlier derivations, however, are only valid for open systems close to equilibrium. Here, the classical limit of the generalized Clausius inequality is valid for closed systems arbitrarily far from equilibrium.

4.2 Generalized Clausius inequality

4.2.3 Upper estimation of the relative entropy

In the last subsection we proposed a lower bound on the irreversible entropy production in terms of the Bures length (4.39). The derivation is based on an analytic, sharp bound on the relative entropy. The relative entropy, however, is a complicated function, which grows exponentially. Hence, we ask for a further upper bound in order to estimate the relative entropy from below and from above. To this end, we make use of various algebraic inequalities. We start with the estimation [OZ01],

$$\text{tr}\{\rho_1 \ln \rho_1 - \rho_1 \ln \rho_2\} \leq \frac{1}{\nu}\text{tr}\{\rho_1^{1+\nu} \rho_2^{-\nu} - \rho_1\}, \tag{4.41}$$

which is true for all positive definite operators, ρ_1 and ρ_2 and $\nu > 0$. For our purpose we can concentrate on the final nonequilibrium and equilibrium densities, ρ_τ and ρ_τ^{eq}. Further, we choose $\nu = 1$ and obtain with the normalization of density operators, $\text{tr}\{\rho\} = 1$, the upper bound,

$$\text{tr}\{\rho_\tau \ln \rho_\tau - \rho_\tau \ln \rho_\tau^{eq}\} \leq \text{tr}\{\rho_\tau^2 (\rho_\tau^{eq})^{-1}\} - 1. \tag{4.42}$$

For the sake of simplicity we further estimate the relative entropy in Eq. (4.42) by making use of the inequality [Mir75],

$$|\text{tr}\{\rho_1 \rho_2\}| \leq \sum_{r=1}^{n} \sigma_r^1 \sigma_r^2, \tag{4.43}$$

which holds for any complex $n \times n$ matrices ρ_1 and ρ_2 with descending singular values, $\sigma_1^1 \geq ... \geq \sigma_n^1$ and $\sigma_1^2 \geq ... \geq \sigma_n^2$. The singular values of an operator T acting on a Hilbert space are defined as the eigenvalues of the operator $\sqrt{T^\dagger T}$. If the ρ_1 and ρ_2 are density operators acting on the same Hilbert space, Eq. (4.43) remains true for arbitrary dimensions and the singular values are identically given by the corresponding eigenvalues [Gri91]. Hence, Eq. (4.42) can be further estimated to yield,

$$\text{tr}\{\rho_\tau \ln \rho_\tau - \rho_\tau \ln \rho_\tau^{eq}\} \leq \sum_n \frac{\left(p_n^0\right)^2}{p_n^\tau} - 1, \tag{4.44}$$

where, as before, $p_n^0 \propto \exp\left(-\beta E_n^0\right)$ and $p_n^\tau \propto \exp\left(-\beta E_n^\tau\right)$ denote initial and final eigenvalues of the equilibrium density operators. Note that the upper bound on the relative entropy in Eq. (4.44) is independent of the nondiagonal matrix elements

75

4 Unitary quantum processes in thermally isolated systems

of ρ_τ and ρ_τ^{eq}. In the general case the quantum relative entropy can, now, be estimated by an only recently proposed inequality [SS00],

$$\sum_{x \in \mathscr{X}} \frac{p^2(x)}{q(x)} - 1 \leq \frac{1}{4}\left(\max_{x \in \mathscr{X}}\left\{\frac{p(x)}{q(x)}\right\} - \min_{x \in \mathscr{X}}\left\{\frac{p(x)}{q(x)}\right\}\right)^2. \qquad (4.45)$$

Hence, the final upper bound on the quantum relative entropy reads,

$$\Delta S_{\text{ir}} \leq \frac{1}{4}\left(\max_n\left\{\frac{p_n^0}{p_n^\tau}\right\} - \min_n\left\{\frac{p_n^0}{p_n^\tau}\right\}\right)^2, \qquad (4.46)$$

which generalizes the earlier published classical inequality [SS00] taking nondiagonal density operators into account. For thermal states, however, where $p_n \propto \exp(-\beta E_n)$, the latter inequality (4.46) is only useful for $E_n^\tau < E_n^0$. Merely in this case the right hand side takes a finite value and infinity otherwise. On the contrary, the irreversible entropy production can always be estimated from above with inequality (4.44). The upper bound can then be evaluated explicitly with the eigenvalues of the according initial and final equilibrium states of the system.

In the latter upper estimation of the relative entropy we rediscovered the algebraic difficulties arising from nondiagonal density operators. Moreover, a physical interpretation of the upper bound (4.46) is lacking. Therefore, we turn in the following to a physically fundamental bound on the maximal rate of entropy production.

4.3 Maximal rate of entropy production

Nonequilibrium irreversible phenomena are not only characterized by the irreversible entropy change, but also by the rate of entropy production, $\sigma = \Delta S_{\text{ir}}/\tau$. The entropy rate, σ, is a central quantity that is associated with the speed of evolution of a nonequilibrium process [dGM84]. In the last chapter we identified a fundamental minimal time, τ_{QSL}, which limits the speed of quantum evolution. It can be written in terms of the time average mean energy, E_τ, and variance, ΔE_τ, as (3.81),

$$\tau_{\text{QSL}} = \max\left\{\frac{\hbar \mathscr{L}(\rho_\tau, \rho_0)}{E_\tau}, \frac{\hbar \mathscr{L}(\rho_\tau, \rho_0)}{\Delta E_\tau}\right\}, \qquad (4.47)$$

4.3 Maximal rate of entropy production

where $\mathscr{L}(\rho_\tau, \rho_0)$ is the Bures length (3.30) between initial, ρ_0, and final state, ρ_τ. Hence, the maximal rate of irreversible entropy production, $\sigma_{\max} = \Delta S_{\text{ir}}/\tau_{\text{QSL}}$, is bounded from above by the minimal time the quantum system needs to evolve from its initial to its final state. For the sake of clarity we, furthermore, write ΔS_{ir} explicitly as,

$$\Delta S_{\text{ir}} = \beta \langle H_\tau \rangle - \beta \langle H_0 \rangle - \beta F_\tau + \beta F_0. \tag{4.48}$$

By making use of a formulation of the second law, $\Delta F = F_\tau - F_0 \leq \langle H_\tau \rangle - \langle H_0 \rangle$, the irreversible entropy production can be estimated from above by,

$$\Delta S_{\text{ir}} \leq 2\beta \left(\langle H_\tau \rangle + \langle H_0 \rangle \right), \tag{4.49}$$

where we, moreover, assumed for the sake of simplicity that $\Delta F \geq 0$. Now, combining Eqs. (4.47) and (4.49) the maximal rate of irreversible entropy production, $\sigma_{\max}$, is given by,

$$\sigma_{\max} \leq \frac{2\beta \left(\langle H_\tau \rangle + \langle H_0 \rangle \right)}{\hbar \mathscr{L}(\rho_\tau, \rho_0)} \min \{E_\tau, \Delta E_\tau\}. \tag{4.50}$$

The latter equation constitutes a fundamental upper bound on the quantum speed of evolution during a nonquasistatic process. The upper bound goes to infinity for identical states, $\rho_\tau = \rho_0$, and takes its finite minimum for orthogonal initial and final states governed by the time averaged energy during the process of the system with respect to its initial and final state. The fundamental maximal rate (4.50) is a mere quantum result, which is based on the Heisenberg uncertainty of energy and time. A classical equivalent is lacking and, hence, the maximal rate is unbounded in classical systems. Moreover, Eq. (4.50) is a generalization of the information-theoretic Bremermann-Bekenstein bound [Bre67, Bek74, Bek81, BS90], which we will derive from Eq. (4.50). The Bremermann-Bekenstein bound is an upper limit on the entropy, S, that can be contained within a given finite region of space which has a finite amount of energy, E. Let us consider large excitations, for which initial and final state become orthogonal, $\mathscr{L}(\rho_0, \rho_\tau) = \pi/2$,

$$\sigma_{\max} \leq \frac{4\beta \left(\langle H_\tau \rangle + \langle H_0 \rangle \right)}{\hbar \pi} \min \{E_\tau, \Delta E_\tau\}. \tag{4.51}$$

Furthermore, we can assume for large excitations and continuous parameterizations that the final energy is much larger than the initial one. For large heating rates, $\langle H_\tau \rangle \gg \langle H_0 \rangle$, we hence conclude,

$$\sigma_{max} \lesssim \frac{4\beta \langle H_\tau \rangle}{\hbar \pi} \min\{E_\tau, \Delta E_\tau\}. \tag{4.52}$$

In the high temperature limit, $\hbar \beta \ll 1$, the latter Eq. (4.52) further simplifies by noting, $E_\tau \simeq 1/\beta$ ($\Delta E_\tau \simeq E_\tau/\sqrt{N} \leq E_\tau$ for N degrees of freedom),

$$\sigma \leq \frac{4}{\hbar \pi} \langle H_\tau \rangle. \tag{4.53}$$

The Bremermann-Bekenstein bound gives the maximum quantum communication rate (capacity) that is possible through a noiseless single channel with signals of finite duration. We stress that the present derivation is solely based on the thermodynamic definition of the entropy production (4.20) and does not make any reference to information entropy or channels; it is, thus, free of the caveats of the original derivations, such as the use of the periodic boundary condition approximation [BS90]. Finally, in the classical limit, $\hbar \to 0$, the bound becomes arbitrarily large and, hence, entropy can be produced in classical systems arbitrarily fast.

4.4 Illustrative example - the parameterized oscillator

In the latter subsection we generalized the Clausius inequality by deriving a sharp lower bound for the irreversible entropy production (cf. Eq. (4.39)). Moreover, we found a fundamental quantum upper bound on the speed of nonequilibrium processes, namely a maximal rate of entropy production (cf. Eq. (4.50)). The present section illustrates the latter results for a completely, analytically solvable system, namely the parameterized harmonic oscillator. As before (cf. Eq. (3.40)), the Hamiltonian reads,

$$H_t = \frac{p^2}{2M} + \frac{M}{2} \omega_t^2 x^2. \tag{4.54}$$

where the angular frequency, ω_t, is varied from an initial value, ω_0, to a final value, ω_1, during the time interval, $0 \leq t \leq \tau$ (cf. [Def08]).

4.4 Illustrative example - the parameterized oscillator

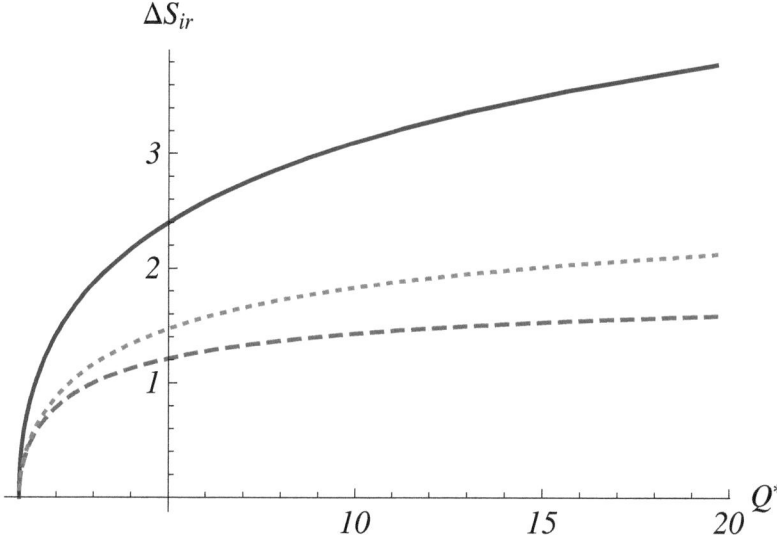

Figure 4.3: Irreversible entropy production (4.55) (solid line) together with the lower bound (4.39) in lowest order expansion (dashed line) and including higher orders (4.38) (dotted line).

4.4.1 Lower bound on entropy production

Now, we want to illustrate the sharpened Clausius inequality (4.39). To this end, we, first, have to evaluate the irreversible entropy production, ΔS_{ir}. For the parameterized harmonic oscillator we obtain [DL08b, AL10] (cf. appendix B),

$$\Delta S_{\text{ir}} = \frac{\beta \hbar (\omega_1 - \omega_0)}{2} Q^* \coth\left(\frac{\beta}{2}\hbar\omega_0\right) - \ln\left(\frac{\sinh(\beta/2\hbar\omega_0)}{\sinh(\beta/2\hbar\omega_1)}\right), \qquad (4.55)$$

where Q^* is again the nonadiabaticity measure (B.11). For the lower bound in terms of the Bures length, $\mathscr{L}$, we need an explicit expression for the fidelity function, F, between the current state, ρ_τ, and its equilibrium counter part, ρ_τ^{eq}. For Gaussian states the fidelity can be evaluated (cf. subsection 3.2.2) and the result is given by the lengthy expression in Eq. (3.47). In Fig. 4.3 we plot the irreversible entropy production, ΔS_{ir}, as a function of Q^* together with the lower bound in

4 Unitary quantum processes in thermally isolated systems

the lowest order expansion and including the first five non-vanishing orders. The farther away from equilibrium the process operates, i.e. the higher the value of Q^*, the more entropy is produced irreversibly. Furthermore, the quadratic order converges quickly to its maximal value, whereas including the higher terms significantly sharpens the estimation.

4.4.2 Maximal rate of entropy production

For the illustration of the maximal rate of entropy production, $\sigma_{\max}$ in Eq. (4.50), we have to evaluate the fidelity between the initial density operator, ρ_0, and the final one, ρ_τ. Analogously to the above derivation (cf. subsection 3.2.2) we obtain,

$$F(\rho_0,\rho_\tau) = \frac{2\left(\cosh(\beta\hbar\omega_0)-1\right)}{\sqrt{3-\mathscr{Q}+(1+\mathscr{Q})\cosh(2\beta\hbar\omega_0)-2}}, \qquad (4.56)$$

where we introduced the quantity,

$$\mathscr{Q} = \frac{1}{2\omega_0^2}\left\{\omega_0^2\left[\omega_0^2 X_\tau^2 + \dot{X}_\tau^2\right] + \left[\omega_0^2 Y_\tau^2 + \dot{Y}_\tau^2\right]\right\}. \qquad (4.57)$$

Similarly to Q^* the quantity $\mathscr{Q}$ is a measure of the adiabaticity of the process. Since, however, the fidelity between the initial equilibrium and the final nonequilibrium state is independent of the equilibrium state of the final configuration, the final frequency, ω_1, does not explicitly appear. Hence, $\mathscr{Q}$ is a mere mathematical tool which lacks a clear physical interpretation in contrast to Q^* (cf. appendix B). However, for cyclic processes, $\omega_1 \equiv \omega_0$, the measure of adiabaticty, Q^*, reduces to $\mathscr{Q}$. Moreover, we evaluate again the low and high temperature limits. In the zero temperature regime, $\hbar\beta \gg 1$, the fidelity $F(\rho_0,\rho_\tau)$ reduces to,

$$F(\rho_0,\rho_\tau)\Big|_{\hbar\beta\gg 1} = \sqrt{\frac{2}{1+\mathscr{Q}}}. \qquad (4.58)$$

On the other hand, we obtain in the classical limit, $\hbar\beta \ll 1$,

$$F(\rho_\tau,\rho_\tau^{eq})\Big|_{\hbar\beta\ll 1} = \frac{2}{1+\mathscr{Q}}. \qquad (4.59)$$

It is remarkable that the zero temperature limit is given by the square root of the classical result. Now, we illustrate in Fig. 4.4 the maximal rate, $\sigma_{\max}$ in Eq. (4.50).

4.5 Experimental realization in cold ion traps

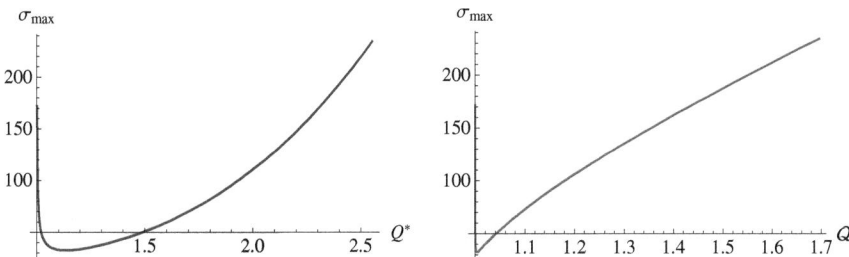

Figure 4.4: Maximal rate of irreversible entropy production, $\sigma_{\max}$, (4.50) as a function of Q^* (B.11) (left) and $\mathcal{Q}$ (4.57) (right).

Here, $\sigma_{\max}$ is plotted as a function of Q^* and $\mathcal{Q}$. For equilibrium processes, $Q^* = \mathcal{Q} = 1$, the maximal rate diverges, since the corresponding processes are either operating infinitely slowly on the system or all intermediate states are almost indistinguishable. For systems driven very far from equilibrium, the maximal rate grows, since the minimal quantum evolution time decreases by investing more energy on the time average. For finite time driving and investing finite energy, the maximal rate of irreversible entropy production has a minimum.

4.5 Experimental realization in cold ion traps

In the preceding chapter as well as in the latter section we illustrated newly derived results with the help of a parameterized harmonic oscillator. Now, the present section is dedicated to an experimental realization in cold ion traps. Thus, we will confirm that the time-dependent harmonic oscillator is not only a toy model, but also has physical relevance. A unique property of ion traps is the possibility to study quantum systems that are either isolated or coupled to tailored quantum environments using reservoir engineering [CZ96, MW00a]. Single ions in radio frequency traps are quantum nanosystems with remarkable properties. They can be laser cooled to very low temperatures, reaching to the motional ground state in the potential. The use of a segmented trap further allows for engineering a vast variety of time-dependent trapping potentials. Hence, trapped ions are not only good candidates for quantum computing, but may also allow us to experimentally approach the verification of the generalized formulations of the second law

4 Unitary quantum processes in thermally isolated systems

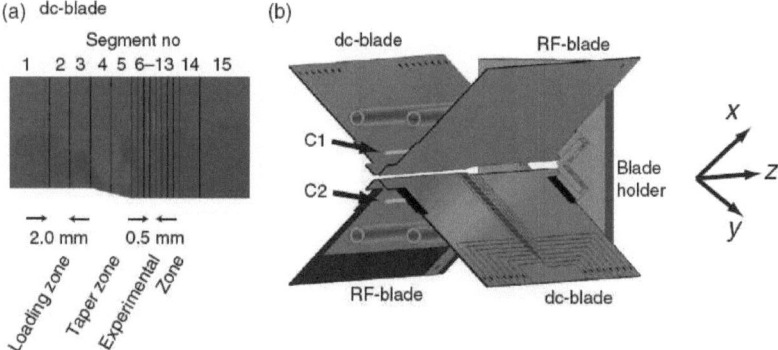

Figure 4.5: Electrode design. (a) Close-up view of the blade design with loading, taper, and experimental zone. (b) Sketch of assembled X-trap consisting of four blades. Compensation electrodes C1 and C2 are parallel to the trap axis. (Taken from [SSK08])

[DL08a].

4.5.1 Experimental set-up

Let us start with a summary of the experimental set-up. For a detailed discussion we refer to [SSK08]. The trap consists of four blades, of which two are connected to a radio frequency (rf) supply and two are segmented with static (dc) voltages, cf. Fig. 4.5. The dc- and rf-blades are assembled perpendicular to each other. The complicated design of the trap is chosen to define the trap potential with high accuracy and to suppress precession of the trapped ion. In Fig. 4.6 we show results of numerical simulations for the confining potential. We observe that in a very good approximation the trap potential is given by the parameterized harmonic oscillator (4.54). The time-dependence of the angular frequency is controlled externally by changing the (dc) voltage at the blades. Besides the external, motional degrees of freedom the ion (e. g. ^{40}Ca$^+$) provides an internal, electronic level scheme. For the further exploration we choose a Λ system comprising the ground state $S_{1/2}$ and two exited states $P_{1/2}$ and $D_{5/2}$. The state $P_{1/2}$ has a short lifetime and decays rapidly into the $S_{1/2}$-state. This decay provides a high spontaneous photon scatter

4.5 Experimental realization in cold ion traps

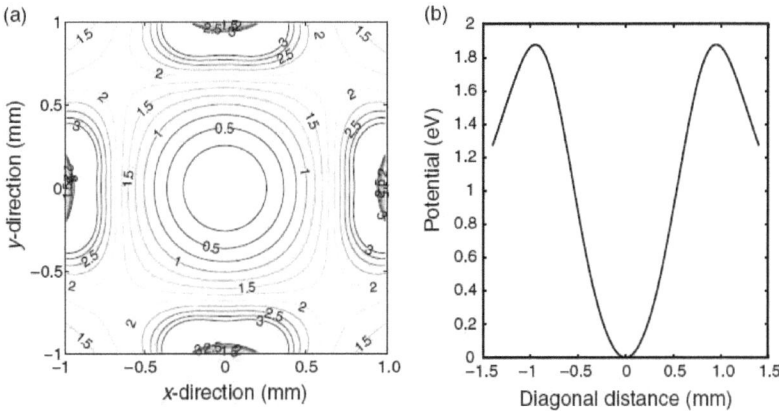

Figure 4.6: (a) Contour plot of the potential in the $(x-y)$-plane in the experimental zone. (b) Cross-section trough potential along $x = y$-direction. (Taken from [SSK08])

rate used for fluorescence detection. The $D_{5/2}$-state is populated with laser pulses, if the spectral bandwidth of the $S_{1/2} - D_{5/2}$ exciting light field is small compared to the sideband structure. Thus, coherent laser pulses allow the exploitation and storage of the motional quantum state information in the internal quantum states.

4.5.2 Verifying the quantum Jarzynski equality

In the beginning of the chapter (cf. subsection 4.1.1) we discussed how to derive the quantum Jarzynski equality (4.7). Now, we describe the experimental verification with the help of cold ion traps introduced above. The measurement procedure can be taken as a paradigm for the experimental realization of other above introduced generalizations of the second law.

The crucial point in experimental verifications is the measurement of the instantaneous quantum state of the system. To this end, a measurement device filtering all possible states is necessary.

83

4 Unitary quantum processes in thermally isolated systems

Filtering scheme

The basic idea of our filtering is a null measurement. To this end, laser pulses are applied deterministically depopulating a single state, which is supposed to be occupied. If a successive fluorescence measurement yields dark, it can be concluded that the system was in the supposed state. In the following, we describe how this idea is implemented for the ion trap.

The *filtering scheme* starts with a sequence of laser pulses being applied to the ion on the narrow S to D transition, coherently processing its internal and external degrees of freedom. This pulse sequence is tailored such that the ion will end in the metastable $D_{5/2}$-state with certainty, if the vibrational quantum state was $|m_{\text{test}}\rangle$. Subsequently, the ion is illuminated with light resonant to the $S_{1/2}$ to $P_{1/2}$ transition. If we observe no fluorescence, the ion is measured in the $D_{5/2}$-state. However, for vibrational states different from $|m_{\text{test}}\rangle$, the laser pulse sequence leads to a superposition state, $\alpha|S_{1/2}\rangle + \beta|D_{5/2}\rangle$, such that there remains a non-vanishing probability $|\beta|^2$ of projecting the superposition into $|D_{5/2}\rangle$, and thus observing no fluorescence. Therefore, the procedure is repeated a few times such that a high quality of the filtering procedure is ensured. Considering the evolution of the quantum state itself, the influence of the above sequence reminds of the operating principle of a filter: its projective *transmission* is unity for a certain input state $|m_{\text{test}}\rangle$ and zero otherwise. This laser pulse sequence is adapted to reach all relevant eigenstates $|n\rangle$ and $|m\rangle$. Moreover, to estimate the time for one whole experiment cycle from preparation to detection, we can assume a few $10\mu\text{sec}$ for sideband pulses and a few hundred μsec fluorescence detection time. Thus, one cycle with multiple filtering iterations will take less than 10 msec, which is short compared to the lifetime of the $D_{5/2}$-state (1.2 sec for $^{40}\text{Ca}^+$) [DL08a]. Therefore, it is assured that the system is fairly isolated from the environment and the dynamics are to very high accuracy unitary. With that filtering scheme the probability distribution of the work (4.1) can be determined and, finally, the quantum Jarzynski equality (4.7) verified.

Measurement protocol

Now, we turn to the experimental *measurement protocol* of the work distribution. It can be summarized to consist of four consecutive steps:

1. The trapped ion is prepared initially in a thermal state with mean phonon

4.5 Experimental realization in cold ion traps

number, $\langle n \rangle = (\exp(\beta \hbar \omega_0) - 1)^{-1}$, in the electronic ground state S-level by laser cooling and optical pumping. To this end, the ion is laser cooled into the vibrational ground state, $|n = 0\rangle$, and subsequently allowed to heat up for a certain time without laser cooling. As the heating rate of the ion within the trap can be precisely measured, this procedure is favorable for very low values of $\langle n \rangle$.

2. In the second step, the initial phonon number, n, is measured using the filtering scheme described in detail above. In this way, the initial energy eigenstate, E_n^0, is determined from spectroscopy measurements.

3. In the third step, the trap potential is varied from an initial value, ω_0, to a final value, ω_τ. This changing potential will, in general modify the ion's motional state into a nonequilibrium state, while its internal electronic state remains unaffected.

4. In the last step, the new phonon number, m, is measured using the filtering scheme and the final energy eigenstate, E_m^τ, is determined. The distribution of the nonequilibrium work, $W = E_m^\tau - E_n^0$ (cf. Eq. (4.1)), is then reconstructed by repeating the measurement sequence. By evaluating both sides of Eq. (4.7) for adiabatic and nonadiabatic processes, the Jarzynski equality can be verified.

The above explained *filtering scheme* and *measurement protocol* are described for the ideal case. However, in real experimental situations one has to handle additional external perturbations. Whereas the filtering and the measurement protocol remain unaffected, the outcome has to be analyzed carefully. Especially, the heating of the cold ion induced by the electronic surroundings might disturb the desired measurement. Hence, we analyze in the following the effect on the work distributions of external perturbations on the ion trap.

4.5.3 Anharmonic corrections and fluctuating electric fields

The two main sources of external perturbations can be identified as imperfections of the harmonic trap and electromagnetic background radiation. Since, however, in the real experiment a main focus lies on avoiding external perturbation, we can

4 Unitary quantum processes in thermally isolated systems

assume anharmonic corrections to the harmonic trap and heating rates to be perturbatively small. Thus, we develop in the present section a perturbative approach to determine the quantum work distribution, $\mathscr{P}(W)$ in Eq. (4.1). We treat in detail the case of a small quartic correction to the potential as well as the effect of a small external fluctuating electric field on a charged harmonic oscillator. Both situations are motivated by the experimental study of the quantum work statistics in linear Paul traps [DL08a]. First, however, we discuss the perturbational approach to the work distribution (4.1) of the harmonic oscillator without external perturbations. Here, we mainly concentrate on the transition probabilities, $p_{m,n}^\tau$, since the initial distributions and the energy eigenvalues are known for the harmonic oscillator for any point of time.

Perturbation theory

To this end, let us start by separating the Hamiltonian (4.54) into the initial harmonic oscillator and a time-dependent part stemming from the change of the angular frequency ω_t,

$$H_t = H_0 + \Omega_t \,, \tag{4.60}$$

where we have introduced the unperturbed, initial Hamiltonian H_0,

$$H_0 = \frac{p^2}{2M} + \frac{M}{2}\omega_0^2 x^2 \,. \tag{4.61}$$

The *perturbation* term in the Hamiltonian is given by,

$$\Omega_t = -\frac{M}{2}\left(\omega_0^2 - \omega_t^2\right) x^2 \,, \tag{4.62}$$

where the latter can be considered as small for small frequency changes. In first order time-dependent perturbation theory, the transition probabilities between initial state, $|n\rangle$, and final state, $|m\rangle$, are given by [DL77b],

$$p_{m,n}^\tau = \left| \delta_{m,n} + \frac{1}{i\hbar} \int_0^\tau dt \, \exp(i\omega_{m,n} t)\, \Omega_{m,n}^t \right|^2 , \tag{4.63}$$

where $\hbar\omega_{m,n} = E_m^0 - E_n^0$ denotes the difference of the unperturbed energy eigenvalues and, $\Omega_{m,n}^t = \langle m|\Omega_t|n\rangle$, are the corresponding interaction matrix elements.

4.5 Experimental realization in cold ion traps

By expressing the position operator, $x = \sqrt{\hbar/2m\omega_0}\,(a^\dagger + a)$, in terms of the usual ladder operators, $a^\dagger|n\rangle = \sqrt{n+1}\,|n+1\rangle$ and $a|n\rangle = \sqrt{n}\,|n-1\rangle$, the interaction matrix elements can be written explicitly as,

$$\Omega^t_{m,n} = -\frac{\hbar}{4\omega_0}\left(\omega_0^2 - \omega_t^2\right)\left[\sqrt{n+1}\sqrt{n+2}\,\delta_{m,n+2}\right.$$
$$\left. + (2n+1)\,\delta_{m,n} + \sqrt{n-1}\sqrt{n}\,\delta_{m,n-2}\right] \quad (4.64)$$

with the Kronecker-delta $\delta_{m,n}$. Equation (4.64) shows that only transitions that satisfy $m = n \pm 2$ are possible. It should be emphasized that this selection rule is at variance with usual textbook examples which contain the selection rule $m = n \pm 1$. The latter applies to a quantum oscillator driven by a small perturbation linear in the position, whereas we here deal with a perturbation (4.62) which is quadratic in x. The full expression of the transition probabilities (4.63) that follow from Eq. (4.64) can be found in [AL10]. In appendix B we propose the completely analytical solution of the parameterized harmonic oscillator. The selection rule, $m = n \pm 2$, is rediscovered in terms of the exact transition probabilities (B.33) and (B.34). The perturbation theory has the advantage that all structural properties of the work distribution are already present for small frequency changes. Now, let us turn to additional external perturbations.

Anharmonic corrections

Above we discussed a method how to experimentally measure the quantum work distribution in modulated ion trap systems. Next, we investigate the influence of a small quartic anharmonicity on the work distribution $\mathscr{P}(W)$. As before, we write the total Hamiltonian as,

$$H_t = H_0 + \Omega_t + A_t, \quad (4.65)$$

where the first anharmonic correction is given by,

$$A_t = \alpha_t x^4. \quad (4.66)$$

The total transition probabilities can then be written as,

$$p^\tau_{m,n} = \left|\delta_{m,n} + \frac{1}{i\hbar}\int_0^\tau dt\,\exp(i\omega_{m,n}t)\,(\Omega^t_{m,n} + A^t_{m,n})\right|^2, \quad (4.67)$$

4 Unitary quantum processes in thermally isolated systems

where $A_{m,n}^t$ are the anharmonic interaction matrix elements. The analytic transition probabilities, $p_{m,n}^\tau$, are again given in [AL10]. The complete results are omitted here due to the lengthy expressions of the formulas. For the sake of clarity, we will continue with a numerical discussion of the structural properties of the resulting work distribution. For the numerical analysis, we choose the parameterization of

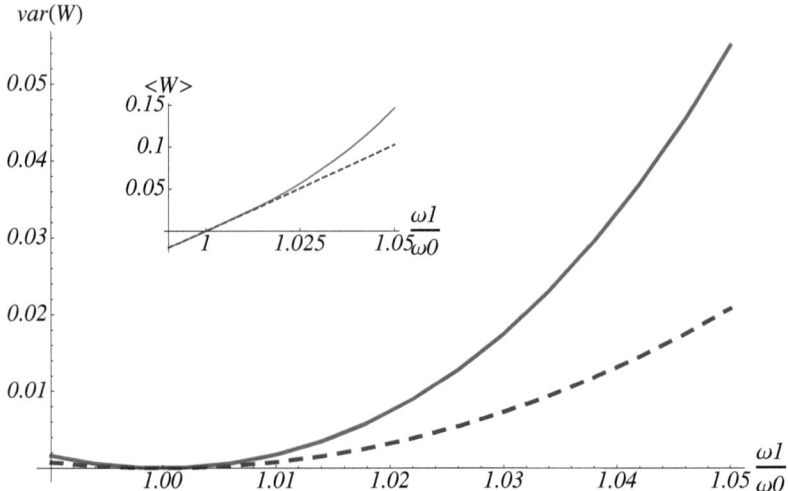

Figure 4.7: Variance and mean (inset) of the work for an oscillator with weak anharmonic corrections (solid line) (4.65) compared with those of the unperturbed oscillator (dashed line) (4.54) ($\omega_0 = 0.5$, $\tau = 1$, $\beta = 0.5$, $M = 1$, $\hbar = 1$ and $\sigma_\alpha = 0.025$).

ω_t^2 to be linear in time,

$$\omega_t^2 = \omega_0^2 + \left(\omega_1^2 - \omega_0^2\right) t/\tau. \qquad (4.68)$$

Since the anharmonic corrections are given by the geometric set-up and, hence, directly scale with the angular frequency of the harmonic oscillator, we assume that

$$\alpha_t = \sigma_\alpha \left(\omega_1^2 - \omega_0^2\right) t/\tau. \qquad (4.69)$$

The parameter σ_α controls the strength of the perturbation. In Fig. 4.7, we have plotted the mean work, $\langle W \rangle$, and the variance, $\text{var}(W) = \langle W^2 \rangle - \langle W \rangle^2$, of the

4.5 Experimental realization in cold ion traps

work distribution for the anharmonically perturbed harmonic oscillator (4.65), together with the exact result for the unperturbed oscillator (4.54). We observe that both quantities are enhanced by the anharmonicity. From the analytical expressions of the transition probabilities (cf. [AL10]), we see that additional transitions $m = n \pm 4$ now become possible because of the quartic correction. These additional transitions lead to a larger mean and variance of the work. Based on our discussion in appendix B, we can, therefore, conclude that the anharmonic perturbation increases the degree of nonadiabaticity of the frequency change. Numerical comparison further shows that the effect of A_t can be neglected up to a strength of roughly one percent, $\sigma_\alpha \lesssim 0.01$, of the harmonic amplitude Ω_t. For a standard trap configuration with trap frequencies of the order kHz-MHz, the harmonic assumption is fulfilled up to energies of the order of eV (cf. [SSK08]) and the effect of anharmonic corrections is negligible for these energies.

Random electric field corrections

Linear Paul traps are almost perfectly isolated from their surroundings. They, however, suffer from the presence of random electric fields that are generated in the trap electrodes [MW00b]. These weak fluctuating fields are the source of motional heating of the charged ions confined in the harmonic trap. The Hamiltonian of the quantum oscillator in the presence of the field is,

$$H_t = H_0 + \Omega_t + \Lambda_t, \tag{4.70}$$

where the small perturbation, Λ_t, is linear in position,

$$\Lambda_t = \lambda_t x. \tag{4.71}$$

The function, $\lambda_t = qE_t$, is proportional to the random electric field E_t (q is the charge of the ion) and is taken to be Gaussian distributed with

$$\langle \lambda_t \rangle = 0 \quad \text{and} \quad \langle \lambda_t \lambda_s \rangle = \kappa_{t,s}. \tag{4.72}$$

The heating rate of the trap is related to the spectral density of the noise λ_t [OT97],

$$\langle \dot{n} \rangle \simeq \frac{1}{4M\hbar\omega_t} \int_{-\infty}^{+\infty} ds \, \exp(i\omega_t s) \langle \lambda_t \lambda_{t+s} \rangle. \tag{4.73}$$

4 Unitary quantum processes in thermally isolated systems

We first calculate the transition probabilities, $p_{m,n}^\tau$, for a fixed value of λ_t and then average over λ_t using Eq. (4.72). In complete analogy to Eq. (4.67), we obtain

$$p_{m,n}^\tau = \left| \delta_{m,n} + \frac{1}{i\hbar} \int_0^\tau dt \, \exp(i\omega_{m,n} t) \left(\Omega_{m,n}^t + \Lambda_{m,n}^t \right) \right|^2, \quad (4.74)$$

with the interaction matrix elements, $\Lambda_{m,n}^t$, given by,

$$\Lambda_{m,n}^t = \lambda_t \sqrt{\frac{\hbar}{2M\omega_0}} \left(\sqrt{n+1}\, \delta_{m,n+1} + \sqrt{n}\, \delta_{m,n-1} \right). \quad (4.75)$$

The explicit expression of the transition probabilities can, again, be found in [AL10]. After averaging over all possible λ_t, the transition probabilities can be divided into two distinct contributions coming from the parametric variation of the frequency (Ω_t in Eq. (4.70)) and the noise term (Λ_t in Eq. (4.70)),

$$\langle p_{m,n}^\tau \rangle_{\lambda_t} = p_{m,n}^\tau(\omega_t) + p_{m,n}^\tau(\langle \lambda_t \lambda_s \rangle). \quad (4.76)$$

Similarly, we can separate the mean final energy into a deterministic and a stochastic part,

$$\langle H_\tau \rangle = \sum_{m,n} E_m^\tau \left(p_{m,n}^\tau(\omega_t) + p_{m,n}^\tau(\langle \lambda_t \lambda_s \rangle) \right) p_n^0$$
$$= \frac{\hbar\omega_1}{2} \left(Q^* + Q_{\lambda_t}^* \right) \coth\left(\frac{\beta}{2} \hbar\omega_0 \right). \quad (4.77)$$

Here, we defined in analogy to the measure of adiabaticty, Q^*, for the unperturbed oscillator the parameter $Q_{\lambda_t}^*$ as,

$$Q_{\lambda_t}^* = \frac{\langle H_\tau \rangle_{\lambda_t}}{\hbar\omega_1/2 \coth(\beta/2\hbar\omega_0)}, \quad (4.78)$$

which is given by the contribution to the internal energy stemming from the external heating,

$$\langle H \rangle_{\lambda_t} = \sum_{n,m} \hbar\omega_1 \left(m + \frac{1}{2} \right) p_{m,n}^\tau(\langle \lambda_t \lambda_s \rangle) p_n^0. \quad (4.79)$$

4.5 Experimental realization in cold ion traps

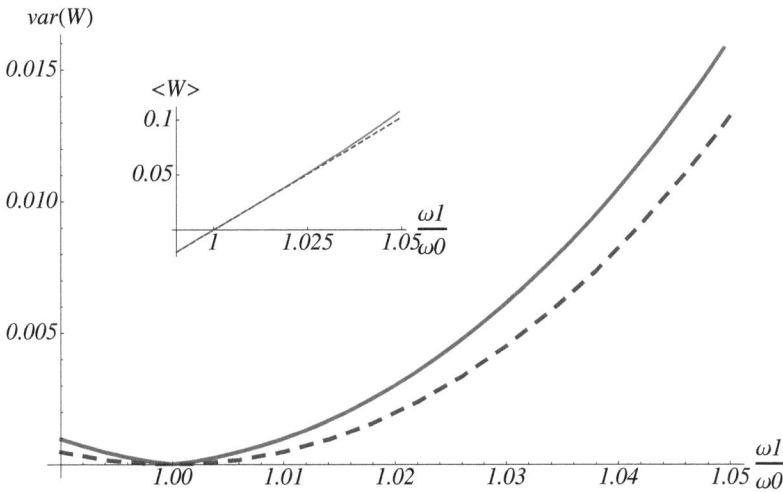

Figure 4.8: Variance and mean (inset) of the work for a charged oscillator with weak electric noise (solid line) (4.65) compared with those of the unperturbed oscillator (dashed line) (4.54) ($\omega_0 = 0.5$, $\tau = 1$, $\beta = 0.5$, $M = 1$, $\hbar = 1$ and $\sigma_\alpha = 0.025$).

Equation (4.77) shows that the effect of the random electric field is to renormalize the adiabaticity parameter $Q^* \to Q^* + Q^*_\lambda$. Both the mean and the variance of the work distribution are increased as depicted in Fig. 4.8. The fluctuating field, thus, enhances the degree of nonadiabaticity. This effect can be understood by noting that the perturbation, Λ_t, generates additional transitions between states (the latter obey $m = n \pm 1$). We observe that the variance is more sensitive to the perturbation than the mean, since it depends quadratically on Q^* and not linearly. For the numerical analysis we have chosen a white noise of the form,

$$\kappa_{t,s} = \sigma_\lambda \left(\omega_0^2 - \omega_1^2 \right) \delta(t-s), \tag{4.80}$$

where the relative noise strength is given by σ_λ. As for the case of the anharmonic perturbation, we note that one can neglect the influence of the electric noise up to a relative strength of roughly one percent, $\sigma_\lambda \lesssim 0.01$.

Finally, we conclude that for small geometric aberrations and weak electromagnetic background radiations effect in apparent more nonadiabatic processes.

4 Unitary quantum processes in thermally isolated systems

Nevertheless, the main structural properties survive as mean and variance remain qualitatively unaffected.

4.6 Summary

The present chapter was dedicated to quantum generalizations of the second law. So far we concentrated on isolated quantum systems, and, hence, unitary dynamics. We started by the proper definition of quantum work and heat pointing out that thermodynamically process dependent quantities are not given by Hermitian operators. Hence, special interest lies on the probability distributions, which lead to various fluctuation theorems. Starting with the work distribution the irreversible entropy production is identified to be given in terms of a relative entropy. This relative entropy serves as the starting point for generalizing and sharpening the Clausius inequality. In particular, we derived a sharp lower bound on the entropy production in terms of the Bures angle between the current nonequilibrium state and its equilibrium counter part. Upper bounds on the relative entropy are more involved and only useful in a limited range of situations. However, we found a fundamental upper bound on the rate of entropy production implied by the quantum speed limit discussed in an earlier chapter 3. This maximal rate is a generalized version of the Bremermann-Bekenstein bound, which can, however, be derived without the caveats of the original derivation. As an illustrative example we discussed in detail a completely analytically solvable system, namely the parameterized harmonic oscillator. Due to its Gaussian properties the entropy production as well as the Bures angle can be evaluated explicitly. The parameterized harmonic oscillator serves as a paradigm for experimental verifications. Especially, linear Paul traps are described with accuracy by isolated harmonic oscillators and we considered how to verify the quantum Jarzynski equality in cold ion traps. Finally, we discussed within a perturbational approach experimental corrections due to geometric aberrations and fluctuating electromagnetic fields. For most experimental situations external perturbations result in an apparent more nonadiabatic process. In conclusion, the present chapter proposed an enlightening set of expressions of the second law for isolated quantum systems arbitrarily far from equilibrium. In the next chapters we will have to deal with the task to generalize the latter results to open systems.

5 Thermodynamics of open quantum systems

In the previous chapter we developed formulations of the second law generalized for isolated quantum systems arbitrarily far from equilibrium. However, real physical systems are always coupled to an environment. Especially in experiments a weak thermal background radiation survives even for the best possible insulation of the set-up. Hence, we discuss in the present chapter various mathematical descriptions and approximations for a quantum system coupled to a heat bath. A lucid introduction to open quantum systems may also be found in the textbook [BP07]. As a new result we will be able to derive a quantum version of a Jarzynski type fluctuation theorem in the weak coupling limit.

5.1 Quantum Langevin equation

As in the classical case (cf. subsection 2.2.1) let us start by considering Langevin dynamics. Starting from a simple model consisting of a particle of interest coupled to an ensemble of harmonic oscillators we will derive the quantum equivalent of the classical Eq. (2.12). However, we will realize that the quantum Langevin equation is mathematically more involved than the classical equivalent. Hence, we will illustrate the physical properties by considering, first, a free Brownian particle, and, second, a particle in a harmonic trap. In particular, we will see that the mean square displacement does not vanish even in the zero temperature limit, which can be interpreted as an expression of the Heisenberg uncertainty relation.

5.1.1 Caldeira-Leggett model

Among the first Ford, Kac, and Mazur [KM65] noted that a thermal environment can be described as an ensemble of coupled harmonic oscillators. This model en-

5 Thermodynamics of open quantum systems

ables to obtain the quantum mechanical form of the Langevin equation (2.12) for a Brownian particle moving in an external potential. In the original derivation the quantum heat bath was given by a chain of coupled harmonic oscillators [KM65]. In the following we consider a simpler model, in which the quantum Brownian particle is coupled to an ensemble of independent, harmonic oscillators [FK87]. The Hamiltonian of the total systems is, then, given by,

$$H = \frac{p^2}{2M} + V(x) + \sum_j \frac{p_j^2}{2M_j} + \frac{1}{2} M_j v_j^2 \left(q_j - \frac{c_j}{M_j v_j^2} x \right)^2, \quad (5.1)$$

where x and p are position and momentum operators of the particle, respectively. The subscript j labels the harmonic oscillators of the bath with position operators q_j, and c_j are the coupling coefficients. The Hamiltonian in Eq. (5.1) is often called the *Caldeira-Leggett model* [CL81, CL83]. Furthermore, to the latter Hamiltonian the canonical commutation relations are appended,

$$[x, p] = i\hbar \quad \text{and} \quad [q_j, p_k] = i\hbar \, \delta_{j,k}, \quad (5.2)$$

where $\delta_{j,k}$ is the Kronecker-delta. The equations of motion for the time-dependent position and momentum operators are given by the corresponding Heisenberg equations (cf. Eq. (2.39)). We obtain for the Brownian particle under consideration,

$$\dot{x} = \frac{p}{M} \quad \text{and} \quad \dot{p} = -V'(x) + \sum_j c_j \left(q_j - \frac{c_j}{M_j v_j^2} x \right), \quad (5.3)$$

and for the jth harmonic oscillator of the bath,

$$\dot{q}_j = \frac{p_j}{M_j} \quad \text{and} \quad \dot{p}_j = -M_j v_j^2 q_j + c_j x. \quad (5.4)$$

The equations of motion (5.4) for the harmonic oscillators mimicking the bath are simple linear differential equations and the solution can be written as,

$$q_j(t) = q_j(0) \cos(v_j t) + \frac{p_j(0)}{M_j v_j} \sin(v_j t) + \frac{c_j}{M_j v_j^2} x_t - \frac{c_j}{M_j v_j^2} x_0 \cos(v_j t)$$
$$- \frac{c_j}{M_j v_j^2} \int_0^t dt' \, \dot{x}_s \cos(v_j (t-s)). \quad (5.5)$$

5.1 Quantum Langevin equation

Substituting Eq. (5.5) in the Heisenberg equations for the particle (5.3) we arrive at,

$$M\ddot{x}_t = -V'(x) - \int_0^t ds\, \Gamma(t-s)\dot{x}_s + \xi_t, \qquad (5.6)$$

where we introduced the force operator, ξ_t, as,

$$\xi_t = \sum_j \left[\left(q_j(0) - \frac{c_j}{M_j v_j^2}x_0\right)\cos(v_j t) + \frac{p_j(0)}{M_j v_j}\sin(v_j t)\right], \qquad (5.7)$$

and where the damping kernel, $\Gamma(t)$, is given by,

$$\Gamma(t) = \sum_j \frac{c_j^2}{M_j v_j^2}\cos(v_j t). \qquad (5.8)$$

Equation (5.6) already has the form of a Langevin equation. Note, however, that Eq. (5.6) is still a deterministic equation. Hence, the stochastic nature of open dynamics has to be introduced with the help of the surrounding harmonic oscillators. To this end, we explicitly make use of the fact that the initial values of the bath, $q_j(0)$ and $p_j(0)$, occur merely in the *external force*, ξ_t; all other terms are functions of the variables of the particle only. Since the harmonic oscillators describe a thermal environment, we assume them to be initially, thermally distributed, $\rho_0 \propto \exp(-\beta H_0)$. With the Hamiltonians of free harmonic oscillators, $H_0 = \sum_j (p_j/2M_j + M_j v_j/2 q_j^2)$ we, thus, obtain,

$$\langle p_j(0) p_j(0)\rangle = \delta_{j,k}\frac{M_j \hbar v_j}{2}\coth\left(\frac{\beta}{2}\hbar v_j\right) \qquad (5.9a)$$

$$\langle \tilde{q}_j(0) \tilde{q}_k(0)\rangle = \delta_{j,k}\frac{\hbar}{2M_j v_j}\coth\left(\frac{\beta}{2}\hbar v_j\right) \qquad (5.9b)$$

$$\langle p_j(0) \tilde{q}_k(0)\rangle = -\langle \tilde{q}_k(0) p_j(0)\rangle = -\frac{1}{2}i\hbar \delta_{j,k}, \qquad (5.9c)$$

where we introduced, $\tilde{q}_j(0) = q_j(0) - c_j/M_j v_j^2 x_0$. In addition, we have the Gaussian property, i.e. the expectation value of an odd number of factors of $q_j(0)$ and $p_j(0)$ vanishes; the expectation value of an even number of the factors is a sum of

5 Thermodynamics of open quantum systems

products of pair expectation values. With the latter correlation functions we, now, obtain the two-time correlation function of the force operator, ξ_t in Eq. (5.7),

$$\frac{1}{2}\langle \xi_t \xi_s + \xi_s \xi_t \rangle = \frac{\hbar}{2} \sum_j \frac{c_j^2}{M_j v_j} \coth\left(\frac{\beta}{2} \hbar v_j\right) \cos\left(v_j(t-s)\right), \quad (5.10)$$

and for the two-time commutator,

$$[\xi_t, \xi_s] = -i\hbar \sum_j \frac{c_j^2}{M_j v_j} \sin\left(v_j(t-s)\right). \quad (5.11)$$

Equation (5.6) together with the statistical properties of the force operators, ξ_t, in Eqs. (5.10) and (5.11) almost constitutes the desired result. However, we still have a single particle coupled to a discrete set of harmonic oscillators, which are thermally distributed. The quantum Langevin equation describes a Brownian particle with a stochastically fluctuating force stemming from a *continuous* bath. Hence, we define the spectral density, $\mathscr{J}(v)$,

$$\mathscr{J}(v) = \sum_j \frac{\pi c_j}{M_j v_j} \delta(v - v_j), \quad (5.12)$$

where $\delta(.)$ is the Dirac-δ-function. With the spectral density, $\mathscr{J}(v)$, we can rewrite the damping kernel, $\Gamma(t)$, the random force, ξ_t, and, thus, Eq. (5.6) continuously. The result is a set of equations constituting the quantum generalization of the classical Langevin equation (2.12) with Gaussian, colored, quantum noise,

$$M \ddot{x}_t = -V'(x) - \int_0^t ds\, \Gamma(t-s)\, \dot{x}_s + \xi_t \quad (5.13a)$$

$$\Gamma(t-s) = \frac{1}{\pi} \int \frac{dv}{v} \mathscr{J}(v) \cos(v(t-s)) \quad (5.13b)$$

$$\frac{1}{2}\langle \xi_t \xi_s + \xi_s \xi_t \rangle = \frac{\hbar}{2\pi} \int dv\, \mathscr{J}(v) \coth\left(\frac{\beta}{2} \hbar v\right) \cos(v(t-s)) \quad (5.13c)$$

$$[\xi_t, \xi_s] = -\frac{i\hbar}{\pi} \int dv\, \mathscr{J}(v) \sin(v(t-s)). \quad (5.13d)$$

Note that the quantum Langevin equation (5.13) is an equation for the position operator, x_t, of the quantum Brownian particle. Moreover, the quantum nature of

5.1 Quantum Langevin equation

the heat bath introduces memory effects as the fluctuating force operator, ξ_t, is not δ-correlated. The classical limit of Gaussian white noise, $\langle \xi_t\xi_s + \xi_s\xi_t\rangle \propto \delta(t-s)$, is rediscovered in the high temperature limit, $\hbar\beta \ll 1$, $\coth(\beta/2\hbar v) \simeq 1$ and for a strictly Ohmic spectral density, $\mathscr{J}(v) = 2M\gamma v$. Due to the complicated correlation function of ξ_t the solution of quantum Langevin dynamics is mathematically much more involved than the classical equivalent. Especially the derivation of fluctuation theorems, or quite generally, formulations of the second law for open quantum systems far from equilibrium obeying Langevin dynamics is still an unsolved problem. Nevertheless, let us further illustrate the properties of the quantum Langevin equation (5.13) for two simple cases before we continue with thermodynamics in the next section.

5.1.2 Free particle

The Langevin equation simplifies for the special case of a free particle, $V(x) \equiv 0$. Moreover, we assume the spectral density, $\mathscr{J}(v)$, to be Ohmic, $\mathscr{J}(v) = 2M\gamma v$, and, thus, we have,

$$M\ddot{x}_t + M\gamma \dot{x}_t = \xi_t \tag{5.14a}$$

$$\frac{1}{2}\langle \xi_t\xi_s + \xi_s\xi_t\rangle = \frac{M\gamma\hbar}{\pi} \int dv\, v\, \coth\left(\frac{\beta}{2}\hbar v\right) \cos(v(t-s)). \tag{5.14b}$$

Note that the correlation function (5.14b) still represents memory effects. Only in the classical limit, $\hbar\beta \ll 1$, the noise correlation (5.14b) reduces to a δ-function, and, thus, white Gaussian noise. However, the latter differential equation (5.14a) can be solved in the Laplace space. The Laplace transform of an arbitrary operator O_t is defined as,

$$\widetilde{O}_\varsigma = \int_0^{+\infty} dt\, \exp(-\varsigma t)\, O_t. \tag{5.15}$$

Then the solution $\widetilde{x}_\varsigma$ of Eq. (5.14b) reads in the Laplace space with initial position, x_0, and initial velocity, v_0,

$$\widetilde{x}_\varsigma = \frac{\widetilde{\xi}_\varsigma + M\gamma x_0 + Mx_0\varsigma + Mv_0}{M\varsigma^2 + M\gamma\varsigma}. \tag{5.16}$$

5 Thermodynamics of open quantum systems

The time-dependent position operator, x_t, which solves the quantum Langevin equation (5.14a), is given by the inverse Laplace transform of $\tilde{x}_\varsigma$. To this end, we introduce an auxiliary Green's function, $\tilde{G}(\varsigma) = 1/(M\varsigma^2 + M\gamma\varsigma)$, whose inverse Laplace transform, $G(t)$, reads [BM90b],

$$G(t) = \frac{1}{M\gamma}\left(1 - \exp(-\gamma t)\right). \tag{5.17}$$

Hence, a solution of the quantum Langevin equation for a free Brownian particle (5.14) can be written as,

$$x_t = x_0 + \frac{v_0}{\gamma}\left(1 - \exp(-\gamma t)\right) + \int_0^t ds\, G(t-s)\,\xi_s. \tag{5.18}$$

As in the classical case (cf. subsection 2.2.1), special interest lies on the mean square displacement and the two-time correlation function of position. The remaining subsection is dedicated to the derivation of closed expressions for the latter two quantities. For the sake of simplicity, we assume the initial velocity to be zero, $v_0 = 0$. We can, then, rewrite the solution x_t as,

$$x_t = x_0 + \int_{-\infty}^t ds\, G(t-s)\,\xi_s - \int_{-\infty}^0 ds\, G(-s)\,\xi_s, \tag{5.19}$$

where the last term in the latter equation can be identified with the initial position, x_0. With the reformulation of the solution, x_t, in Eq. (5.19), the integrals appearing in the two-time correlation function, $\frac{1}{2}\langle x_t x_s + x_s x_t\rangle$, can be evaluated,

$$\frac{1}{2}\langle x_t x_s + x_s x_t\rangle = \frac{1}{2}\int_0^t ds \int_0^{t'} ds'\, G(t-s)G(t'-s')\,\langle \xi_s \xi_{s'} + \xi_{s'} \xi_s\rangle. \tag{5.20}$$

After a few lines of calculation we obtain,

$$\frac{1}{2}\langle x_t x_s + x_s x_t\rangle = \frac{\hbar\gamma}{\pi M}\int_0^\infty dv\, \frac{\coth(\beta/2\hbar v)}{v(\gamma^2 + v^2)} \\ \times \left(\cos(v(t-t')) - \cos(vt) - \cos(vt') + 1\right), \tag{5.21}$$

5.1 Quantum Langevin equation

and for the mean square displacement by taking $t = s$,

$$\left\langle x_t^2 \right\rangle = \frac{2\hbar\gamma}{\pi M} \int_0^\infty dv \, \frac{\coth(\beta/2\hbar v)}{v(\gamma^2 + v^2)} (1 - \cos(vt)). \quad (5.22)$$

Note that the angular brackets $\langle ... \rangle$ denote an ensemble average over the fluctuations of the quantum heat bath. The latter expressions (5.21) and (5.22) are still given in terms of spectral integrals. However, in the classical limit, $\hbar\beta \ll 1$, the integral in Eq. (5.22) can be evaluated and the mean square displacement can be written in closed form,

$$\left\langle x_t^2 \right\rangle \simeq \frac{4\gamma}{\pi M \beta} \int_0^\infty dv \, \frac{1 - \cos(vt)}{v^2(\gamma^2 + v^2)} = \frac{2}{\beta M \gamma} \left[t + \frac{1}{\gamma} (\exp(-\gamma t) - 1) \right], \quad (5.23)$$

which coincides with the usual, classical mean square displacement of a free Brownian particle [Ris89]. On the other hand, in the zero temperature limit, $\hbar\beta \gg 1$, we obtain,

$$\begin{aligned}\left\langle x_t^2 \right\rangle &\simeq \frac{2\hbar\gamma}{\pi M} \int_0^\infty dv \, \frac{1 - \cos(vt)}{v(\gamma^2 + v^2)} \\ &= \frac{2\hbar}{\pi M \gamma} \left[c + \ln(t\gamma) - \cosh(t\gamma)\mathrm{chi}(t\gamma) + \sinh(t\gamma)\mathrm{shi}(t\gamma) \right],\end{aligned} \quad (5.24)$$

where $c = 0.577...$ is the Euler constant. Moreover, chi(.) and shi(.) denote the cosine and sine integral functions, respectively [AS72]. The latter result is mere quantum peculiarity. Even in the zero temperature limit, where thermal fluctuations vanish, the mean square displacement is finite. In other words, even in the zero temperature limit the quantum particle maintains a non-vanishing, statistical width, and, hence, Eq. (5.24) can be understood as an expression of the Heisenberg uncertainty relation for position and momentum. A detailed analysis of the long time behavior of Eq. (5.24) can also be found in [SI87] and an approximate discussion in [FO01].

5.1.3 Harmonic potential

In the latter subsection we analyzed the quantum Langevin dynamics of a free, Brownian particle. For arbitrary potentials the solution of the quantum Langevin

5 Thermodynamics of open quantum systems

equation (5.13) cannot be written in closed form. However, for the harmonic oscillator the integration becomes feasible. Thus, we consider in the present subsection a quantum Brownian particle in a harmonic trap, $V(x) = M/2\,\omega^2 x^2$, and, as before, an Ohmic spectral density, $\mathscr{J}(v) = 2M\gamma v$. Thus, the Langevin equation (5.13) takes the form,

$$M\ddot{x}_t + M\gamma \dot{x}_t + M\omega^2 x = \xi_t \tag{5.25a}$$

with correlation function

$$\frac{1}{2}\langle \xi_t \xi_s + \xi_s \xi_t \rangle = \frac{M\gamma\hbar}{\pi} \int dv\, v \coth\left(\frac{\beta}{2}\hbar v\right) \cos\left(v(t-s)\right). \tag{5.25b}$$

As before, the quantum Langevin equation (5.25a) can be solved in the Laplace space and a solution is given by,

$$\tilde{x}_\varsigma = \frac{\tilde{\xi}_\varsigma + Mx_0\varsigma + Mv_0 + M\gamma x_0}{M\varsigma^2 + M\gamma\varsigma + M\omega^2}, \tag{5.26}$$

where again x_0 and v_0 are initial position and velocity of the Brownian particle, respectively. The Green's function, $\widetilde{G}(\varsigma) = 1/(M\varsigma^2 + M\gamma\varsigma + M\omega^2)$, takes the form,

$$G(t) = \frac{2\exp(-\gamma t/2)\sinh\left(t/2\sqrt{\gamma^2 - 4\omega^2}\right)}{M\sqrt{\gamma^2 - 4\omega^2}}. \tag{5.27}$$

Analogously to Eq. (5.19), we can rewrite the solution, x_t, of Eq. (5.25a) with zero initial velocity, $v_0 = 0$, in terms of the Green's function, $G(t)$, as,

$$x_t = \int_{-\infty}^{t} ds\, G(t-s)\, \xi_s. \tag{5.28}$$

Then the two-time correlation function results in,

$$\begin{aligned}\frac{1}{2}\langle x_t x_s + x_s x_t \rangle &= \frac{1}{2}\int_0^t ds \int_0^{t'} ds'\, G(t-s)G(t'-s') \langle \xi_t \xi_s + \xi_s \xi_t \rangle \\ &= \frac{\hbar\gamma}{\pi M} \int_0^\infty dv\, \frac{v \coth(\beta/2\hbar v)}{(\omega^2 - v^2)^2 + \gamma^2 v^2} \\ &\quad \times \left(\cos\left(v(t-t')\right) - \cos(vt) - \cos\left(vt'\right) + 1\right),\end{aligned} \tag{5.29}$$

and the mean square displacement reads,

$$\langle x_t^2 \rangle = \frac{2\hbar\gamma}{\pi M} \int_0^\infty dv \, \frac{v \coth(\beta/2\hbar v)}{(\omega^2 - v^2)^2 + \gamma^2 v^2} (1 - \cos(vt)). \qquad (5.30)$$

In contrast to the free particle $\langle x_t^2 \rangle$ remains even in the classical as well as in the zero temperature limit a non-trivial integral. Hence, one might conclude that the dynamics of open quantum systems are mathematically more involved than in the classical case.

In particular for driven systems it is not feasible to formulate a simple description universally valid for arbitrary potentials. In general, neither in terms of Langevin equations nor with the help of quantum master equations (cf. section 5.3) dynamical properties of open, time-dependent quantum systems are completely analytically analyzable.

5.2 Thermodynamics in the weak coupling limit

Since we saw in the last section that the dynamics of open quantum systems might become mathematically difficult, the present section is dedicated to a purely thermodynamic analysis of open quantum systems. Let us start by considering a quantum system of interest, which is weakly coupled to a thermal environment. The Hamiltonian of the total system, H, can, then, be separated into system, H_S, bath, H_B, and an interaction term, h_γ (cf. Eq. (4.8)),

$$H = H_t^S \otimes \mathbb{1}^B + \mathbb{1}^S \otimes H^B + h_\gamma. \qquad (5.31)$$

In Fig. 5.1 we illustrate the physical situation of a subsystem of interest separated from its thermal surroundings. We allow the Hamiltonian of the system of interest, H_t^S, to be time-dependent. Hence, work is performed on the system as the Hamiltonian varies with an external work parameter. Moreover, for the sake of simplicity, we assume in the present section the interaction Hamiltonian, $\langle h_\gamma \rangle$, to be small compared to the mean energy of the system, $\langle H_t^S \rangle$, for all times. Therefore, the Hilbert subspaces of the system and the bath factorize and effects of the

5 Thermodynamics of open quantum systems

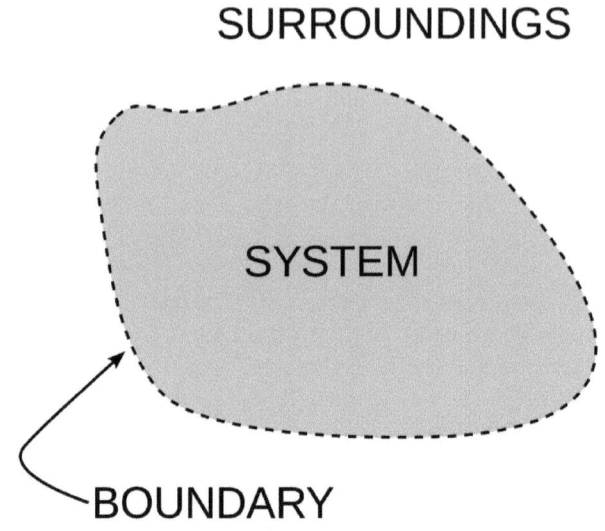

Figure 5.1: Defined quantum system and thermal surroundings

interaction can be treated by means of perturbation theory [BP07]. In the following, however, we propose a completely thermodynamic treatment. To this end, we derive a closed expression for the entropy generated during a process starting in an arbitrary initial state. Furthermore, we will show that this entropy production fulfills a Jarzynski type fluctuation theorem. It is worth emphasizing that the present treatment is independent of caveats arising from Lindblad master equations or quantum trajectories [EM06]. A brief introduction to quantum master equations will be provided in the next section.

5.2.1 Quantum entropy production

Let us begin by considering the first law of thermodynamics in its reduced formulation for the system,

$$\Delta U^S = \left\langle Q^S \right\rangle + \left\langle W^S \right\rangle, \tag{5.32}$$

5.2 Thermodynamics in the weak coupling limit

where ΔU^S denotes the change in the internal energy, $\langle W^S \rangle$ is the work performed during a process acting on the system of interest, and $\langle Q^S \rangle$ the heat exchanged with the environment. With the balance equation (cf. Eq. (2.3)) representing the second law we conclude for the irreversible entropy production,

$$\Delta S_{\text{ir}} = \Delta S^S - \beta \Delta E^S + \beta \langle W^S \rangle . \tag{5.33}$$

In the following we consider processes starting at time $t = 0$ and ending at $t = \tau$. Hence, we denote the change in internal energy by, $\Delta E^S = E^S_\tau - E^S_0$, and the change in entropy by, $\Delta S^S = S^S_\tau - S^S_0$. The change in thermodynamic entropy, ΔS, can be identified with the change in the von Neumann entropy,

$$S^S = -\text{tr}\left\{\rho^S \ln \rho^S\right\} , \tag{5.34}$$

if the initial and final state are close enough to an equilibrium state. As before, we denote by ρ^S the reduced density operator of the system of interest only. Otherwise, however, the system is allowed to visit nonequilibrium states arbitrarily far from thermal equilibrium. The latter is consistent with processes described by means of conventional thermodynamics, where processes are completely determined by their initial and final equilibrium states. Now, by further introducing the instantaneous equilibrium density operator for the reduced system, $\rho_t^{S,\text{eq}} = \exp\left(\beta H^S_t\right)/Z^S_t$, in terms of the time-dependent Hamiltonian, H^S_t, the internal energy, $U_t = \langle H^S_t \rangle$, is written as,

$$\beta U^S_t = -\text{tr}\left\{\rho^S_t \ln \rho_t^{S,\text{eq}}\right\} - \ln Z^S_t . \tag{5.35}$$

Moreover, we used in the latter Eq. (5.35) the definition of the reduced partition function, $Z^S_t = \text{tr}\left\{\exp\left(\beta H^S_t\right)\right\}$. Finally, regarding the last term on the right hand side of Eq. (5.33), the work is identified with the change of the Hamiltonian with time,

$$\beta \langle W^S \rangle = \beta \int_0^\tau dt\, \text{tr}\left\{\rho^S_t \partial_t H^S_t\right\} = -\int_0^\tau dt\, \text{tr}\left\{\rho^S_t \partial_t \ln \rho_t^{S,\text{eq}}\right\} - \ln Z^S_\tau + \ln Z^S_0 , \tag{5.36}$$

5 Thermodynamics of open quantum systems

which we again expressed in terms of the corresponding equilibrium density operator, $\rho_t^{S,eq}$. In the latter Eq. (5.36) we used the conventional identification [Ali02],

$$dU = d\langle H \rangle = \text{tr}\{(d\rho)H\} + \text{tr}\{\rho\,(dH)\} = \langle \delta Q \rangle + \langle \delta W \rangle. \tag{5.37}$$

Now, we conclude by combining Eqs. (5.34)-(5.36) to Eq. (5.33) our main result for the irreversible entropy production,

$$\Delta S_{\text{ir}} = S\left(\rho_0^S \| \rho_0^{S,eq}\right) - S\left(\rho_\tau^S \| \rho_\tau^{S,eq}\right) - \int_0^\tau dt\,\text{tr}\left\{\rho_t^S\,\partial_t \ln \rho_t^{S,eq}\right\}, \tag{5.38}$$

where we again introduced the relative entropy $S(.\|.)$ (cf. appendix A.1). Equation (5.38) indicates three contributions to the irreversible entropy production, ΔS_{ir}. The first term measures the distinguishability between the initial density operator, ρ_0^S, and its equilibrium counter part, $\rho_0^{S,eq}$. In the following we will see that $S(\rho_0^S\|\rho_0^{S,eq})$ is the entropy produced during a relaxation process from the initial nonequilibrium state, ρ_0^S, to equilibrium. The second term measures analogously the remaining entropy production due to the system not relaxing from the final nonequilibrium state, ρ_τ^S, to equilibrium. It has to be subtracted since, generally, systems undergoing arbitrary processes do not end in an equilibrium state. The last term, finally, is the contribution of the irreversible work, $\langle W_{\text{ir}}^S \rangle = \langle W^S \rangle - \Delta F^S$. We denote by $\langle W^S \rangle$ the average total work performed during time τ and ΔF^S is the according free energy difference. In Fig. 5.2 we illustrate the dynamics of the quantum system in a phase space sketch. The upper (blue) line represents a path in phase space where the system is always in a nonequilibrium state. The lower (red) line is the equilibrium analog. The dashed lines connecting the initial and final states, respectively, are the distances between equilibrium and nonequilibrium quantified by the according relative entropies.

Instantaneous rate

The upper result in Eq. (5.38) is obtained by merely use of thermodynamic arguments, only. It can be alternatively derived, by identifying the instantaneous rate first. For the quantum equivalent of the classical balance equation (2.24)

$$\dot{S}^S = \beta \dot{Q}^S + \sigma, \tag{5.39}$$

5.2 Thermodynamics in the weak coupling limit

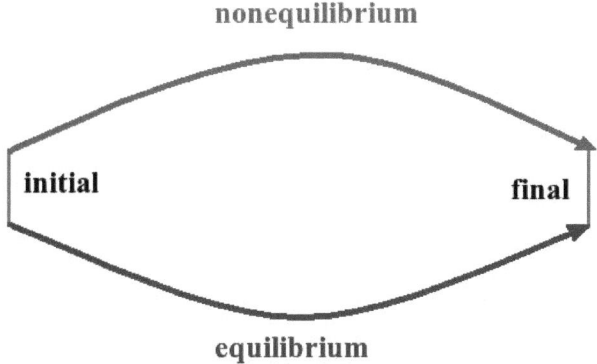

Figure 5.2: Phase space sketch illustrating the dynamics of the system; equilibrium (lower) and nonequilibrium (upper) path.

Spohn derived for time-independent Hamiltonians [Spo78] the instantaneous rate, σ_t, which was generalized by Breuer [Bre03] for time-dependent systems. The result reads (cf. subsection 5.3.1),

$$\sigma_t = -\text{tr}\left\{\left(\partial_t \rho_t^S\right) \ln \rho_t^S\right\} + \text{tr}\left\{\left(\partial_t \rho_t^S\right) \ln \rho_t^{S,\text{eq}}\right\}. \tag{5.40}$$

However, the derivation of the preceding expression in Eq. (5.40) is based on the validity of Lindblad master equations, which we will discuss in the next section. Moreover, the rate of entropy production, σ_t, can be rewritten in terms of the relative entropy,

$$\sigma_t = -\partial_t S\left(\rho_t^S \| \rho_t^{S,\text{eq}}\right) - \text{tr}\left\{\rho_t^S \partial_t \ln \rho_t^{S,\text{eq}}\right\}. \tag{5.41}$$

By integrating the rate, σ_t, over the process time interval form 0 to τ, $\Delta S_{\text{ir}} = \int_0^\tau dt\, \sigma_t$, we rediscover the above proposed expression for the irreversible entropy production, ΔS_{ir} in Eq. (5.38). Independently of our upper analysis, the latter relation has earlier been realized by Lindblad [Lin83]. It is worth emphasizing once again that the validity of the identification of σ_t is given by the applicability

5 Thermodynamics of open quantum systems

of a Lindblad master equation. The upper derivation in the present subsection 5.2.1 is less restrictive, since it is merely based on thermodynamic arguments.

5.2.2 Particular processes

We, now, turn to the discussion of particular quantum processes, in order to clarify the physical meaning of the contributions to ΔS_{ir}. Let us start with a time-independent Hamiltonian and, thus, $\rho_t^{S,\text{eq}} \equiv \rho_0^{S,\text{eq}}$. The latter implies that no work is performed. In a first step we shall be interested in a complete relaxation process, i.e the system starts in an arbitrary nonequilibrium state, ρ_0^S, and reaches after time τ its equilibrium state, $\rho_\tau^S = \rho_0^{S,\text{eq}}$. Hence, the general entropy production (5.38) reduces to,

$$\Delta S_{\text{ir}} = S\left(\rho_0^S \| \rho_0^{S,\text{eq}}\right). \tag{5.42}$$

The latter is identical to the expression derived by Schlögl almost fifty years ago, which was, however, published in a German paper [Sch66, Sch89]. As mentioned earlier, irreversible entropy production is due to the distinguishability between the equilibrium and nonequilibrium density operator measured by the relative entropy, $S\left(\rho_0^S \| \rho_0^{S,\text{eq}}\right)$. If we consider a partial relaxation process, i.e. the system starts in a nonequilibrium state, ρ_0, but does not reach equilibrium in time τ, the entropy production reads,

$$\Delta S_{\text{ir}} = S\left(\rho_0^S \| \rho_0^{S,\text{eq}}\right) - S\left(\rho_\tau^S \| \rho_\tau^{S,\text{eq}}\right). \tag{5.43}$$

The latter indicates the additive character of the entropy. The irreversible entropy production in a partial relaxation process is given by the contribution of a relaxation from the real final state to equilibrium subtracted from the contribution of the complete relaxation.

Next let us consider time-dependent systems. In conventional thermodynamics one usually considers processes where the system starts, $\rho_0^S = \rho_0^{S,\text{eq}}$, and ends, $\rho_\tau^S = \rho_\tau^{S,\text{eq}}$, in equilibrium. In general, the initial and final equilibrium states are not identical. Nevertheless the irreversible entropy production reduces to,

$$\Delta S_{\text{ir}} = -\int_0^\tau dt\, \text{tr}\left\{\rho_t^S \partial_t \ln \rho_t^{S,\text{eq}}\right\}, \tag{5.44}$$

5.2 Thermodynamics in the weak coupling limit

which is nothing else than the irreversible part of the work, $\Delta S_{\text{ir}} = \beta \langle W_{\text{ir}}^S \rangle$. Hence, we conclude that the expression of the entropy production in Eq. (5.38) is the general expression including all possible thermodynamic processes operating on a quantum system weakly coupled to a thermal environment.

From the latter analysis of the entropy production for open quantum system far from equilibrium we conclude that the thermodynamic properties are described in terms of relative entropies. Hence, for nonequilibrium processes the relative entropy between the instantaneous state and a corresponding equilibrium state is the crucial quantity. The thermodynamic information is, therefore, contained in the relative entropy rather than in the von Neumann entropy of the nonequilibrium state.

Unitary dynamics

Finally, we have to prove that Eq. (5.38) is consistent with the expression derived for isolated quantum systems (4.27) starting in an equilibrium state. For unitary dynamics, where the von Neumann entropy is constant, $S_t^S \equiv S_0^S$, the entropy production, ΔS_{ir}, (5.38) can be further evaluated to read,

$$\Delta S_{\text{ir}} = -\int_0^\tau dt \, \text{tr} \left\{ \rho_t^S \, \partial_t \ln \rho_t^{S,\text{eq}} \right\} = \int_0^\tau dt \, \partial_t S \left(\rho_t^S || \rho_t^{S,\text{eq}} \right) = S \left(\rho_\tau^S || \rho_\tau^{S,\text{eq}} \right). \tag{5.45}$$

The latter is identical to the above result in Eq. (4.27), and, thus, the earlier identification of the entropy production as the irreversible part of the work is confirmed.

5.2.3 Jarzynski type fluctuation theorem

In the remaining part of the section we turn to a formulation of the second law. As usual, we like to have a Jarzynski type fluctuation theorem being valid arbitrarily far from equilibrium. To this end, we will, now, identify the entropy production for a single realization of the process and, then, construct the corresponding probability distribution.

5 Thermodynamics of open quantum systems

Fluctuating entropy

So far we considered the irreversible entropy production in the framework of thermodynamics. Our above obtained result (5.38) is the mean entropy production averaged over all possible realizations of a particular process, $\Delta S_{\text{ir}} = \langle \Sigma \rangle$. By Σ we denote the entropy production for a single realization, which is, hence, a fluctuation quantity. By definition $\langle \Sigma \rangle$ is the sum of the change in entropy and the heat transferred to the environment. By making explicit use of the separability of the Hilbert subspaces in the weak coupling limit [BZ06, BP07], the entropy production can be written as,

$$\langle \Sigma \rangle = -\text{tr}\left\{\left(\rho_0^S \otimes \rho^B\right) U_\tau^\dagger \left(\ln \rho_\tau^S \otimes \mathbb{1}^B\right) U_\tau\right\} + \text{tr}\left\{\left(\rho_0^S \otimes \rho^B\right) \left(\ln \rho_0^S \otimes \mathbb{1}^B\right)\right\}$$
$$- \beta Q^S. \tag{5.46}$$

Generally, the last term in Eq. (5.46) is more complicated due to contributions of the change in internal energy and work performed. However, by neglecting contributions of the interaction energy, h_γ, the heat transferred from the system S equals minus the heat absorbed by the bath B, $Q^S \simeq -Q^B$ (cf. subsection 4.1.2). Since no work is performed on the bath the change in internal energy of the bath is identical to the heat exchanged with the system of interest, $\Delta E^B = Q^B$. Thus, the last term in Eq. (5.46) can be written as an average over the same density operator $\rho_0^S \otimes \rho^B$,

$$Q^B = \text{tr}\left\{\left(\rho_0^S \otimes \rho^B\right) U_\tau^\dagger \left(\mathbb{1}^S \otimes H^B\right) U_\tau\right\} - \text{tr}\left\{\left(\rho_0^S \otimes \rho^B\right) \left(\mathbb{1}^S \otimes H^B\right)\right\}. \tag{5.47}$$

The latter Eqs. (5.46) and (5.47) clarify that the entropy production, $\langle \Sigma \rangle$, is a quantity averaged over the initial state, $\rho_0^S \otimes \rho^B$, and quantum transitions induced by the time-dependent contributions. Analogous expressions are obtained for the work distribution in isolated systems (4.1) and the heat exchange between two subsystems (4.12).

Probability distribution

Hence, we postulate the probability distribution of Σ to read,

$$\mathscr{P}(\Sigma) = \sum_{\iota,\phi} p_\iota \langle \iota | U_\tau^\dagger | \phi \rangle \langle \phi | U_\tau | \iota \rangle \, \delta\left(\Sigma - \Sigma_{\iota \to \phi}\right). \tag{5.48}$$

5.2 Thermodynamics in the weak coupling limit

The latter equation generalizes Eq. (4.1) to open systems. Furthermore, we introduced the eigenbasis of the initial density operator, $\rho_0 = \rho_0^S \otimes \rho^B = \sum_\iota p_\iota |\iota\rangle\langle\iota|$, and of the final one, $\rho_t = U_\tau \left(\rho_0^S \otimes \rho^B\right) U_\tau^\dagger = \sum_\phi p_\phi |\phi\rangle\langle\phi|$. Here, we make again explicit use of the weak coupling limit in the sense that we assume the initial and final basis sets as disjoint unions of the subspaces S and B, $|\iota\rangle \in \{|\iota^S\rangle\} \dot\cup \{|\iota^B\rangle\}$ and $|\phi\rangle \in \{|\phi^S\rangle\} \dot\cup \{|\phi^B\rangle\}$. The remaining task is to identify the realization dependent entropy production $\Sigma_{\iota \to \phi}$. To this end, we rewrite Eq. (5.46) with (5.47) by inserting the basis sets $\{|\iota\rangle\}$ and $\{|\phi\rangle\}$ to yield,

$$\langle \Sigma \rangle = - \sum_{\iota^S, \phi^S} p_\iota \ln p_\phi \langle \iota | U_\tau^\dagger | \phi \rangle \langle \phi | U_\tau | \iota \rangle + \sum_{\iota^S, \phi^S} p_\iota \ln p_\iota \langle \iota | U_\tau^\dagger | \phi \rangle \langle \phi | U_\tau | \iota \rangle \\ + \sum_{\iota^B, \phi^B} p_\iota \beta E_\phi^B \langle \iota | U_\tau^\dagger | \phi \rangle \langle \phi | U_\tau | \iota \rangle - \sum_{\iota^B, \phi^B} p_\iota \beta E_\iota^B \langle \iota | U_\tau^\dagger | \phi \rangle \langle \phi | U_\tau | \iota \rangle . \quad (5.49)$$

In Eq. (5.49) we introduced the initial and final energy eigenvalues of the bath, βE_ι^B and βE_ϕ^B, respectively. They can be introduced here, since the bath stays in its thermal equilibrium state, $\rho^B \propto \exp\left(-\beta H^B\right)$, at all times. As usual, U denotes the unitary time evolution operator of the total system. Finally, the mean entropy production can be written as,

$$\langle \Sigma \rangle = \sum_{\iota,\phi} p_\iota \langle \iota | U_\tau^\dagger | \phi \rangle \langle \phi | U_\tau | \iota \rangle \left(\delta_{\iota,\iota^S} \ln p_\iota - \delta_{\phi,\phi^S} \ln p_\phi + \delta_{\phi,\phi^B} \beta E_\phi^B - \delta_{\iota,\iota^B} \beta E_\iota^B \right), \quad (5.50)$$

where we used the Kronecker-delta $\delta_{m,n}$. Further, by comparing the mean of the probability density in Eq. (5.48), $\int d\Sigma \Sigma \mathscr{P}(\Sigma)$, with the average $\langle \Sigma \rangle$ (5.49) we identify the microscopic entropy production for a transition from state $|\iota\rangle$ to a state $|\phi\rangle$ to read,

$$\Sigma_{\iota \to \phi} = \delta_{\iota,\iota^S} \ln p_\iota - \delta_{\phi,\phi^S} \ln p_\phi + \delta_{\phi,\phi^B} \beta E_\phi^B - \delta_{\iota,\iota^B} \beta E_\iota^B. \quad (5.51)$$

The realization dependent entropy production, $\Sigma_{\iota \to \phi}$ in Eq. (5.51), is the quantum mechanical generalization of the trajectory dependent entropy production introduced in the context of stochastic thermodynamics [Sei05]. It includes explicitly contributions of the entropy change in the system of interest and the heat flow to the environment. It is worth emphasizing that $\Sigma_{\iota \to \phi}$ in Eq. (5.51) has been derived by merely making use of the thermodynamically evaluated entropy production, ΔS_ir in Eq. (5.38). At no point the use of a master equation or a *quantum trajectory* has been necessary in contrast to [EM06].

5 Thermodynamics of open quantum systems

Fluctuation theorem

In the remaining paragraph we will show that the above defined entropy production (5.51) indeed fulfills an integrated fluctuation theorem in its common form,

$$\langle \exp(-\Sigma) \rangle = 1. \tag{5.52}$$

To this end, we evaluate $\langle \exp(-\Sigma) \rangle$ with the probability distribution (5.48) and the microscopic entropy production (5.51). Hence, we have

$$\langle \exp(-\Sigma) \rangle = \sum_{\iota,\phi} p_\iota \langle \iota | U_\tau^\dagger | \phi \rangle \langle \phi | U_\tau | \iota \rangle \exp\left(-\delta_{\iota,\iota^S} \ln p_\iota + \delta_{\phi,\phi^S} \ln p_\phi\right) \\ \times \exp\left(-\delta_{\phi,\phi^B} \beta E_\phi^B + \delta_{\iota,\iota^B} \beta E_\iota^B\right), \tag{5.53}$$

which can be split with respect to the subspaces by evaluating the Kronecker-deltas to yield,

$$\langle \exp(-\Sigma) \rangle = \sum_{\iota^S,\phi^S} p_\iota \langle \iota | U_\tau^\dagger | \phi \rangle \langle \phi | U_\tau | \iota \rangle \exp\left(-\ln p_\iota + \ln p_\phi\right) \\ + \sum_{\iota^B,\phi^B} p_\iota \langle \iota | U_\tau^\dagger | \phi \rangle \langle \phi | U_\tau | \iota \rangle \exp\left(-\beta E_\phi^B + \beta E_\iota^B\right). \tag{5.54}$$

Since the bath B remains in thermal equilibrium at its initial, $p_{\iota^B} \propto \exp\left(-\beta E_\iota^B\right)$, as well as at its final state, $p_{\phi^B} \propto \exp\left(-\beta E_\phi^B\right)$, we simplify Eq. (5.54) to read,

$$\langle \exp(-\Sigma) \rangle = \left(\sum_{\iota^S,\phi^S} + \sum_{\iota^B,\phi^B}\right) p_\phi \langle \iota | U_\tau^\dagger | \phi \rangle \langle \phi | U_\tau | \iota \rangle = \text{tr}\{\rho_\tau\} = 1, \tag{5.55}$$

which concludes the proof of Eq. (5.52). In Eq. (5.55) we, moreover, used the normalization of the probability distributions. Consequently, we derived the quantum generalization of an integrated fluctuation theorem for systems initially in arbitrary nonequilibrium states. To this end, we explicitly made use of the separability of the Hilbert subspaces as given by the weak coupling limit. The latter fluctuation theorem (5.52) is a quantum version of the classical general theorems derived by Seifert in [Sei08].

It is worth emphasizing once again that the above derivations are completely based on thermodynamic arguments. Hence, the fluctuation theorem (5.52) is valid for all kinds of processes starting and ending close to an equilibrium state. Nevertheless, the quantum system may evolve arbitrarily far from equilibrium at intermediate instants. This is in complete agreement with thermodynamics, where processes are fully characterized by their initial and final equilibrium state. Therefore, our approach does not rely on the formulation of master equations for driven quantum systems as it was done in earlier considerations [EM06].

5.3 Statistical physics of open quantum systems

In the beginning of the present chapter, we introduced the quantum Langevin equation. The latter turned out to be a non-trivial stochastic, operator equation with colored noise. In the last section, we derived a closed expression for the entropy production and proved a Jarzynski type fluctuation theorem. In the classical case (cf. chapter 2) we saw that an equivalent description of Langevin dynamics becomes possible in terms of probability distributions for the microstate. Starting from the Fokker-Planck description we were able to derive the fluctuation theorems, as well. Thus, it would be desirable to obtain a quantum analog of that approach. However, we will see in the present section that the quantum equivalent of Fokker-Planck dynamics is physically as well as mathematically involved. In particular the definition of work and heat is still an unsolved problem. However, in the next chapter we will be able to derive the quantum entropy production and a fluctuation theorem in the strong coupling regime. In order to deepen the insight into open quantum dynamics, the present section briefly analyzes quantum equivalents of the classical Klein-Kramers equation for various couplings between the system of interest and the environment. In quantum mechanics probability distributions translate to density operators. Hence, we shall be interested in the evolution equation of the reduced density operator, $\rho^S = \mathrm{tr}_B\{\rho\}$, the *quantum master equation*,

$$i\hbar\, \mathrm{d}_t \rho_t^S = \left[H_t^S, \rho_t^S\right] + \mathfrak{D}\left(\rho_t^S\right), \qquad (5.56)$$

which is the von Neumann equation for the reduced system with an additional contribution in terms of the superoperator, $\mathfrak{D}$, which describes the interaction with the environment. The dissipative term in the quantum master equation (5.56) denoted by $\mathfrak{D}$ is responsible for the reduced dynamics not being unitary. Hence, we focus

5 Thermodynamics of open quantum systems

on expressions of $\mathfrak{D}$ in the following considerations. In the present section, we concentrate on the physical discussion and interpretation of the resulting master equations, whereas we refer to the literature for their derivations [BP07]. We start with the Markovian approximation in the weak coupling limit, before we turn to more involved equations allowing for non-perturbational small interactions with the environment. Moreover, we will restrict ourselves in the present discussion to time-independent Hamiltonians since merely for the harmonic oscillator the actual form of $\mathfrak{D}$ has been clarified for time-dependent problems [ZH95].

5.3.1 Markovian approximation

In an earlier section 5.1 we derived and discussed the quantum Langevin equation. We saw that in the general case a quantum mechanical heat bath is characterized by colored noise. Colored noise, however, is an expression of memory effects, and, thus, processes described by quantum Langevin dynamics are non-Markovian. Markov processes, on the other hand, are mathematically well-understood. Therefore, the question arises, whether under certain circumstances the dynamics of an open quantum system can be approximated by a Markov process. An answer was given by Lindblad [Lin74, Lin75, Lin76], who proposed the *Lindblad master equation*,

$$i\hbar d_t \rho_t^S = \left[H^S, \rho_t^S\right] + \sum_k \gamma_k \left(A_k \rho_t^S A_k^\dagger - \frac{1}{2} A_k^\dagger A_k \rho_t^S - \frac{1}{2} \rho_t^S A_k^\dagger A_k\right), \quad (5.57)$$

where the A_k are called the *Lindblad operators*. The latter master equation (5.57) can be derived from microscopic principles [BP07] and we concentrate, here, on the properties and conditions of validity.

Properties of Lindblad dynamics

Let us summarize the most important properties of the dynamics generated by Eq. (5.57):

- The linear map, $\mathscr{F}$, which is given in terms of the Lindblad operators, A_k,

$$d_t \rho_t^S = \mathscr{F} \rho_t^S, \quad (5.58)$$

5.3 Statistical physics of open quantum systems

is the generator of the quantum dynamical semigroup [Ali02]. Furthermore, the semigroup property of a formal solution, $\mathscr{S}(t) = \exp(\mathscr{F}t)$, of Eq. (5.58) is the generalization of a Markov condition to open quantum dynamics,

$$\mathscr{S}(t_1)\mathscr{S}(t_2) = \mathscr{S}(t_1 + t_2) \quad \forall t_1, t_2 \geq 0. \tag{5.59}$$

- The Lindblad master equation (5.57) is the most general, time-homogeneous equation describing the evolution of the reduced density operator, ρ_t^S, which is trace preserving and completely positive for any initial condition.

- The non-negative quantities γ_k have the dimension of an inverse time provided the A_k are taken to be dimensionless. It can be shown [BP07] that the γ_k are given in terms of correlation functions of the thermal environment and play the role of relaxation rates for different decay modes of the open system.

- The stationary solution of Eq. (5.57) is a Boltzmann-Gibbs equilibrium distribution, $\rho_{\text{stat}}^S \propto \exp(-\beta H^S)$, which is encoded in the construction of the Lindblad operators, A_k,

$$\left[H^S, A_k A_k^\dagger\right] = \left[H^S, A_k^\dagger A_k\right] = 0 \quad \forall k. \tag{5.60}$$

- The form of Eq. (5.57) is not unique, since the set of Lindblad operators is closed under unitary transformations.

Physical applicability of the model

Sometimes the Lindblad master equation (5.57) is supposed to be unphysical, since it completely ignores the quantum nature of the heat bath. A quantum heat bath is characterized by its quantum memory (cf. Eq. (5.13)), i.e. non-Markovian dynamics. However, in e.g. quantum optics, where the systems of interest are fairly decoupled from the surroundings, the Lindblad equation becomes physically applicable. In an earlier discussion (cf. section 4.5) we considered a cold ion trap. These ion traps are only locally cooled down to their motional ground state. That means that merely the ion reaches very low temperatures, whereas the surroundings remain unaffected. Hence, the dynamics of an *open* ion trap can be understood as an ultra-cold quantum system coupled to a classical environment. This situation is perfectly described by the Lindblad master equation (5.57).

5 Thermodynamics of open quantum systems

Entropy production

As noted earlier Spohn derived the rate of entropy production for Lindblad dynamics [Spo78]. Starting again with the balance equation,

$$\sigma_t = \dot{S} - \beta \langle \dot{Q} \rangle, \qquad (5.61)$$

and further identifying, $\dot{S} = -\mathrm{tr}\{\mathfrak{D}(\rho_t^S)\ln\rho_t^S\}$, and, $\langle \dot{Q} \rangle = \mathrm{tr}\{\mathfrak{D}(\rho_t^S)H^S\}$, we obtain for the rate of entropy production, σ_t,

$$\sigma_t = -\mathrm{tr}\{\mathfrak{D}(\rho_t^S)\ln\rho_t^S\} - \beta\,\mathrm{tr}\{\mathfrak{D}(\rho_t^S)H^S\}. \qquad (5.62)$$

Note that σ_t is governed by the dissipative superoperator $\mathfrak{D}$, since the unitary part of the time evolution can be absorbed by the cyclic invariance of the trace. From a physical point of view, the rate of entropy production, σ_t, is completely determined by the interaction of the system of interest and its surroundings. The irreversible entropy is the part of the total entropy increase, which flows into the heat bath, and, thus, cannot be re-obtained by the system [dGM84]. For isolated systems, and, hence, unitary dynamics, the irreversible entropy is the entropy, which would flow to the environment, if the system was coupled to a heat bath. So far the expression for the rate of entropy production in Eq. (5.62) is completely general. For Lindblad dynamics, where the stationary state is given by thermal equilibrium state, we can write,

$$\sigma_t = -\mathrm{d}_t S\left(\rho_t^S \| \rho_{\mathrm{eq}}^S\right), \qquad (5.63)$$

where we again used the relative entropy, $S(.\|.)$. The latter expression (5.63) for σ_t is consistent with our earlier thermodynamic discussion (cf. Eq. (5.41)) for time-dependent systems. Moreover, Spohn proved in [Spo78] that σ_t in Eq. (5.63) is positive for all times.

5.3.2 Quantum Brownian motion

In the preceding subsection we presented the Lindblad master equation, which arises in the so called quantum optical limit. The physical conditions underlying the approximations are that the systematic evolution of the reduced system is fast compared with a typical relaxation time. However, in various physical situations involving strong system-environment coupling and low temperatures this

5.3 Statistical physics of open quantum systems

condition is violated. In contrast to quantum optics it may even happen that the dynamics of the system of interest is slow compared to the correlation times of the environment. Under such circumstances the Lindblad master equation (5.57) is not applicable and another scheme of approximation becomes necessary. In the present subsection we are concerned with the Caldeira-Leggett master equation [CL83], whose derivation starts with the microscopic model (5.1). As we saw earlier, one of the crucial quantities in quantum Brownian motions is the spectral density, $\mathscr{J}(v)$. In a phenomenological modelling one often deals with an Ohmic damping, $\mathscr{J}(v) = 2M\gamma v$. However, the high frequency modes of the environment lead to a renormalization of the physical parameters in the particle potential [BP07]. To account for this renormalization one introduces a high-frequency cutoff, Ω. An Ohmic spectral density with e.g. a Lorentz-Drude cutoff reads,

$$\mathscr{J}(v) = 2M\gamma v \frac{\Omega^2}{\Omega^2 + v^2}. \tag{5.64}$$

We can, now, determine the typical time scales in terms of the high-frequency cutoff, Ω.

Correlation time approximations

It is easy to see that with an Ohmic spectral density with cutoff Ω the largest correlation time, τ_B, is given by [BP07],

$$\tau_B = \max\left\{\frac{1}{\Omega}, \frac{\hbar\beta}{2\pi}\right\}. \tag{5.65}$$

Similarly, to a possible microscopic derivation of the Lindblad master equation (5.57) one assumes the coupling to be weak,

$$\hbar\gamma \ll \max\left\{\hbar\Omega, \frac{2\pi}{\beta}\right\}, \tag{5.66}$$

for which a Born-Markov approximation becomes possible. The latter condition (5.66) corresponds to $\tau_B \ll \tau_R$, where τ_R is the typical relaxation time, with $\tau_R = 1/\gamma$. However, for the applicability of the Lindblad master equation one also assumes that a typical time scale of the reduced system, τ_S, is short compared to the typical relaxation time, τ_R. By contrast, we are, here, interested in the case

5 Thermodynamics of open quantum systems

where the evolution of the reduced system is slow in comparison to the bath correlation time. With a typical frequency of the system, $\omega_0 = 1/\tau_S$, the latter condition becomes,

$$\hbar\omega_0 \ll \min\left\{\hbar\Omega, \frac{2\pi}{\beta}\right\}. \tag{5.67}$$

The main difference between Lindblad dynamics and the present model can be summarized as: for the validity of Eq. (5.57) a typical time scale of the system has to be much less than the relaxation time, $\tau_S \ll \tau_R$. For the present model, which is taking account of quantum effects of the bath, we assume that a typical correlation of the bath is much shorter than the time scales of the reduced system, $\tau_B \ll \tau_S$.

Caldeira-Leggett master equation

With the latter conditions (5.66) and (5.67) a master equation of the general form (5.56) can be deduced. After a lengthy derivation [CL83, BP07] one obtains,

$$i\hbar\, d_t \rho_t^S = \left[H^S, \rho_t^S\right] + \gamma\left[x, \left\{p, \rho_t^S\right\}\right] - \frac{2iM\gamma}{\hbar\beta}\left[x, \left[x, \rho_t^S\right]\right], \tag{5.68}$$

where $\{.,.\}$ is the anti-commutator. As usual the first term on the right hand side of the master equation describes the free, unitary evolution of the system of interest. The second term, which is proportional to the damping coefficient γ, is a dissipative term. Finally, the last term describes thermal fluctuations and it can be shown that it is of fundamental importance for *decoherence*. Furthermore, the classical limit of the master equation (5.68) in position representation reduces to the above discussed Klein-Kramers equation (2.23). It is worth mentioning, however, that in the general case the stationary solution of Eq. (5.68) is not given by a Boltzmann-Gibbs distribution, of which one easily convinces oneself by substituting $\rho \propto \exp\left(-\beta H^S\right)$ in Eq. (5.68).

Entropy production

Coming back to thermodynamics, let us evaluate the rate of entropy production (5.62) for the Caldeira-Leggett master equation (5.68). With the help of the com-

5.3 Statistical physics of open quantum systems

mutator expressions,

$$[x, H^S] = \frac{i\hbar}{M} p \tag{5.69a}$$

$$[x^2, H^S] = \frac{i\hbar}{M} \{p, x\} \tag{5.69b}$$

we obtain after a few lines of calculation,

$$\sigma_t = \text{tr}\left\{ \left([\ln \rho_t^S, x] + \frac{\beta i\hbar}{M} p\right) \left(\frac{i\gamma}{\hbar}\{p, \rho_t^S\} + \frac{2M\gamma}{\beta \hbar^2}[x, \rho_t^S]\right)\right\}. \tag{5.70}$$

The latter result in Eq. (5.70) is the quantum generalization of the classical rate of entropy production (2.31) under the conditions (5.66) and (5.67). Further evaluation of σ_t is only feasible for particular systems, since the rate of entropy production (5.70) is governed by the eigenstates of the reduced density operator, ρ_t^S.

5.3.3 Hu-Paz-Zhang master equation

The upper master equations (5.57) and (5.68) are approximate results, which are applicable under certain conditions and physical circumstances. As we saw for the quantum Langevin equation (5.13) the general case is mathematically involved. However, for the harmonic oscillator (cf. subsection 5.1.3) the situation simplifies a lot. Hu, Paz, and Zhang [PZ92] proposed the exact master equation for a harmonic oscillator coupled to a bath of harmonic oscillators, which reads,

$$\begin{aligned}i\hbar d_t \rho_t^S &= \left[H^S, \rho_t^S\right] + \Gamma_t\left[x, \{p, \rho_t^S\}\right] - iM\Gamma_t h_t\left[x, [x, \rho_t^S]\right] \\ &\quad - i\Gamma_t f_t\left[x, [p, \rho_t^S]\right] - \frac{1}{2}M\delta\omega_t^2\left[x^2, \rho_t^S\right].\end{aligned} \tag{5.71}$$

The time-dependent coefficients Γ_t, h_t, f_t and $\delta\omega_t$ are non-trivial functions of the spectral density, $\mathscr{J}(\nu)$, and the coupling to the environment. The first term of the right hand side of the Hu-Paz-Zhang master equation (5.71) is the dissipative term with a time-dependent dissipative constant Γ_t. The third and fourth terms are the diffusive terms with time-dependent coefficients $\Gamma_t h_t$ and $\Gamma_t f_t$, respectively. The last term, finally, depicts a time-dependent frequency shift $\delta\omega_t^2$. The explicit time dependence of these coefficients is rather complicated, however, given a particular form of the spectral density, $\mathscr{J}(\nu)$, and the initial state of the environment,

5 Thermodynamics of open quantum systems

they can be calculated. The master equation (5.71) describes the non-Markovian behavior of a quantum Brownian particle in a harmonic potential. Finally, for a strictly Ohmic spectral density and high temperatures, $\hbar\beta \ll 1/\Omega$, with the high frequency cutoff, Ω, the Caldeira-Leggett master equation (5.68) is recovered. It is worth noting that the Hu-Paz-Zhang master equation was generalized to harmonic oscillators with time-dependent angular frequency by Zerbe and Hänggi [ZH95].

5.4 Summary

The present chapter was dedicated to a thermodynamic description of open quantum systems. We started by introducing the Caldeira-Leggett model, which led to the derivation of the quantum Langevin equation. The quantum Langevin equation is a stochastic operator equation with colored, Gaussian, quantum noise. Hence, the classical derivations of generalizations of the second law are not completely transferable to open quantum dynamics. Therefore, we turned to a thermodynamics approach to systems in the weak coupling limit. For such situations we derived a closed expression of the total entropy production during a processes operating arbitrarily far from equilibrium. We found that the crucial thermodynamic quantity for open, nonequilibrium quantum systems is the relative entropy between the instantaneous state and a corresponding equilibrium one. Apparently, the relative entropy contains the thermodynamic information in contrast to the instantaneous von Neumann entropy during an arbitrary process. Furthermore, we were able to derive a Jarzynski type fluctuation theorem for the entropy production. Our derivation is independent of Lindblad dynamics or hypothetic quantum trajectories. It is solely based on thermodynamic arguments and in analogy to the treatment of isolated quantum systems. Finally, we briefly introduced the quantum master equations. A quantum master equation is the quantum analog of the classical Fokker-Planck description of Brownian motion. However, the resulting equations turned out to be either mathematically involved or only applicable under a restrictive set of conditions. Hence, a thermodynamic analysis of open quantum systems starting with a general master equation for the reduced dynamics is still an unsolved problem.

6 Strong coupling limit - a semiclassical approach

In the previous chapter we introduced mathematical descriptions of the dynamic properties of open quantum systems. We found that quantum Brownian motion is much more involved than its classical analog. The quantum Langevin equation is a stochastic operator equation with colored noise, and the quantum master equations are only applicable in the weak friction regime. Hence, derived generalizations of the second law, and, particularly, closed expressions for the entropy production are merely valid for weak enough coupling of the system to its thermal surroundings. To deepen the insight in thermodynamics of open quantum systems we, now, turn to the opposite simplifying limit. The present chapter is dedicated to the strong friction regime. First, we will briefly introduce the quantum Smoluchowski equation [PG01, TM07, Tu04, DL09], before we, second, derive a detailed as well as an integral fluctuation theorem. A notable advantage of the following description is that the usual classical definitions of work and heat are valid in contrast to the full quantum regime [LH07]. Finally, we will propose a Josephson junction experiment that would allow to test our predictions.

6.1 Quantum Smoluchowski dynamics

Let us start by briefly reviewing the main properties of the approximations in the high friction regime. First, we will present the path integral formalism for the reduced system dynamics. Then, we discuss various approximations and their range of validity, before we, finally, introduce the quantum Smoluchowski equation. We will see that the description by means of the quantum Smoluchowski equation is a semiclassical approach, where quantum effects manifest themselves as additional quantum fluctuations. The meaning of these quantum fluctuations will be elucidated with the analytic expression for the escape rate from metastable wells. The

6 Strong coupling limit - a semiclassical approach

present discussion follows roughly [JAP05], where we will correct several results (cf. also [AG08]).

6.1.1 Reduced dynamics in path integral formulation

As before we consider a total quantum system, which can be separated into a system of interest, H^S, a thermal environment, H^B, and an interaction term, h_γ,

$$H = H^S \otimes \mathbb{1}^B + \mathbb{1}^S \otimes H^B + h_\gamma. \tag{6.1}$$

The heat bath is mimicked by an ensemble of independent harmonic oscillators bilinearly coupled to the system (cf. Caldeira-Leggett model in Eq. (5.1)). By properly eliminating the bath degrees of freedom one obtains the dissipative dynamics of the reduced system. In contrast to the above discussion (cf. chapter 5) the interaction term, h_γ, is not perturbatively small, but rather governs the dynamics of the reduced system. The quantum dynamics of the reduced system follows from,

$$\rho_t^S = \text{tr}_B \left\{ \exp\left(-iHt/\hbar\right) \rho_0 \exp\left(iHt/\hbar\right) \right\}, \tag{6.2}$$

where ρ_0 is the initial state of the total system. In the conventional Feynman-Vernon theory [FV63] it is assumed that the initial state, ρ_0, can be factorized. This is one of the assumptions appearing in the last chapter 5 and can be interpreted as the system and bath being isolated from each other at $t = 0^-$. In the strong friction regime, however, the initial state has to be correlated [SI88]. For the sake of simplicity, we assume the initial state to read in the initial position representation with coordinates x_i,

$$\langle x_i | \rho_0^S | x_i' \rangle = \langle x_i | \rho_\beta^S | x_i' \rangle \Lambda \left(x_i, x_i' \right), \tag{6.3}$$

where $\langle q | \rho_\beta^S | x' \rangle = \text{tr}_B \left\{ \langle x | \exp\left(-\beta H\right) | x' \rangle \right\}$ and the multiplicative term $\Lambda(x_i, x_i')$ is the position representation of a preparation function. The reduced dynamics (6.1) with an initial state (6.3) can, now, be described within the Feynman path integral formalism. Since the thermal environment is modeled as an ensemble of harmonic oscillators, it can be integrated and one obtains in the final position representation with coordinates x_f,

$$\langle x_f | \rho_t^S | x_f' \rangle = \int dx_i \int dx_i' J\left(x_f, x_f', t, x_i, x_i'\right) \Lambda\left(x_i, x_i'\right), \tag{6.4}$$

6.1 Quantum Smoluchowski dynamics

where the propagating function $J(.)$ is a threefold path integral over the the system degree of freedom only. With the reduced partition function of the system only, $Z^S = \text{tr}\{\exp(-\beta H)\}/Z^B$, the integral kernel, $J(.)$, can be written as,

$$J\left(x_f, x'_f, t, x_i, x'_i\right) = \frac{1}{Z^S} \int \mathscr{D}[x] \int \mathscr{D}[x'] \int \mathscr{D}[\bar{x}] \exp\left(-\frac{1}{i\hbar}\mathfrak{S}[x, x', \bar{x}]\right). \quad (6.5)$$

The two real time paths $x(s)$ and $x'(s)$ connect the initial points x_i and x'_i with the fixed end points x_f and x'_f, respectively. On the contrary, $\bar{x}(\varsigma)$ is an imaginary time path, which runs from x_i to x'_i in the interval $\hbar\beta$. The contribution of each path is weighted with an effective action, $\mathfrak{S}[x, x', \bar{x}] = S^S[x] - S^S[x'] + i\overline{S}^S[\bar{x}] + i\Phi[x, x', \bar{x}]$. Hence, the effective action consists of the actions of the reduced system in real and imaginary time and an additional interaction contribution Φ. The latter one is completely determined by the damping kernel $\Gamma(t)$ (5.8). So far no approximations have been applied and Eq. (6.5) is generally valid. Due to the mathematical difficulties of further evaluation we turn, now, to the limit of strong coupling of the reduced system to its environment.

6.1.2 Quantum strong friction regime

The classical Klein-Kramers equation reduces to the Smoluchowski equation if the time scales of fast equilibration of momentum and slow equilibration of position separate (cf. subsection 2.2.2). Moreover, for dissipative quantum systems one expects that friction makes the system of interest to behave more classically due to decoherence effects. In particular, the complicated path integral expression (6.5) should simplify considerably in the limit of strong damping. The latter expectation can be shown rigorously to hold [PG01] and we continue with the main results and the range of validity of the approximations. As in the discussion of the Caldeira-Leggett master equation (5.68) we assume an Ohmic spectral density with Lorentz-Drude cutoff (cf. Eq. (5.64)),

$$\mathscr{J}(\nu) = 2M\gamma\nu \frac{\Omega^2}{\Omega^2 + \nu^2}. \quad (6.6)$$

6 Strong coupling limit - a semiclassical approach

Then, the strong damping limit can be expressed with a typical frequency ω_0 of the reduced system (e.g. its ground state frequency) as,

$$\frac{\gamma}{\omega_0^2} \gg \frac{\hbar\beta}{2\pi}, \frac{1}{\Omega}, \frac{1}{\gamma}. \qquad (6.7)$$

Note that the latter conditions (6.7) express the opposite limit as the separation of time scales in Eqs. (5.66) and (5.67). Here, the separation of time scales implying the classical Smoluchowski equation (2.32) must additionally incorporate the time scale of quantum fluctuations, $\hbar\beta$. The idea is now, to evaluate for strong friction, $\gamma/\omega_0 \gg 1$, the path integral expression (6.4) on a coarse-grained time scale, $s \gg \hbar\beta, 1/\Omega, 1/\gamma$. In this approximation we observe:

- Nondiagonal elements of the reduced density matrix are strongly suppressed during the time evolution.

- The real time part of the kernel $\Gamma(t)$ (5.8) becomes local on the coarse-grained time scale.

- Initial correlations described by Eq. (6.3) survive for times of order γ/ω_0^2 verifying that the initial state can be assumed to factorize.

Applying the latter recipe to the path integral expression in Eq. (6.4) the quantum Smoluchowski equation can be derived as the evolution equation of the diagonal elements $\langle x_f | \rho_t^S | x_f \rangle$.

6.1.3 Quantum Smoluchowski equation

In the semiclassical picture introduced in the previous subsection, the quantum system follows a *classical trajectory* and quantum effects manifest themselves through *quantum fluctuations* that act in addition to the thermal fluctuations induced by the heat bath. Then, according to the latest publications [DL09, MA10a, MA10b], the quantum Smoluchowski equation can be written as,

$$\partial_t p(x,t) = \frac{1}{\gamma M} \partial_x \left[V'(x) + \frac{1}{\beta} \partial_x D_e(x) \right] p(x,t), \qquad (6.8)$$

6.1 Quantum Smoluchowski dynamics

where $V'(x)$ is the derivative of the external potential with respect to position and β the inverse temperature. The effective diffusion coefficient $D_e(x)$ is given by,

$$D_e(x) = \frac{1}{1 - \lambda \beta V''(x)}, \qquad (6.9)$$

where we introduced the quantum parameter λ. It measures the size of the quantum fluctuations and can be evaluated explicitly for the spectral density (6.6) to read,

$$\lambda = \frac{\hbar}{\pi \gamma M} \left[c + \Psi \left(\frac{\gamma \hbar \beta}{2\pi} + 1 \right) \right]. \qquad (6.10)$$

Here, $c = 0.577...$ is the Euler constant and Ψ the digamma function [AS72]. It should be noted that quantum corrections depend explicitly on the position of the system through the curvature of the potential $V''(x)$. When $\lambda = 0$, Eq. (6.8) reduces to the classical Smoluchowski equation (5.8) with constant diffusion coefficient. Moreover, in the high temperature limit, $\gamma \hbar \beta \ll 1$, the quantum parameter can be expressed in leading order as $\lambda \simeq \hbar^2 \beta^2 / 12M$, which coincides with the equilibrium quantum correction proposed earlier by Wigner [Wig32]. Of particular interest, however, is the limit of strong damping, $\gamma \hbar \beta \gg 1$, where $\lambda \simeq \hbar / M \gamma \pi \ln(\gamma \hbar \beta / 2\pi)$. Thus, the quantum Smoluchowski approach is beyond perturbation theory and for strong friction quantum systems can be described over a wide range of temperatures, $1/\beta$. Finally, one can conclude that off-diagonal matrix elements, $\langle x | \rho_t^S | x' \rangle$, are strongly suppressed by friction and of the order $\mathcal{O}(1/\sqrt{\gamma \ln(\Omega/\gamma)})$ [JAP05].

The stationary equilibrium solution of Eq. (6.8), with natural boundary conditions, is

$$p_s(x) = \frac{1}{Z^S} \exp\left(-\beta V(x) + \lambda \beta^2 V'(x)^2 / 2 \right) \left[1 - \lambda \beta V''(x) \right], \qquad (6.11)$$

where Z^S is again the partition function of the reduced system. It is worth emphasizing that the above equilibrium expression is in general non-Gibbsian when $\lambda \neq 0$, which is in accordance with the above discussion. Once again we emphasize that the quantum Smoluchowski equation (6.8) with the effective diffusion coefficient (6.9) is valid in the semiclassical range of parameters, $\gamma/\omega_0^2 \gg (\hbar \beta, 1/\gamma)$, $\hbar \gamma \gg 1/\beta$ and $|\lambda \beta V''(x)| < 1$, where ω_0 is a characteristic frequency, i.e. the curvature at a potential minimum of the system. In order to deepen our insight, we

6 Strong coupling limit - a semiclassical approach

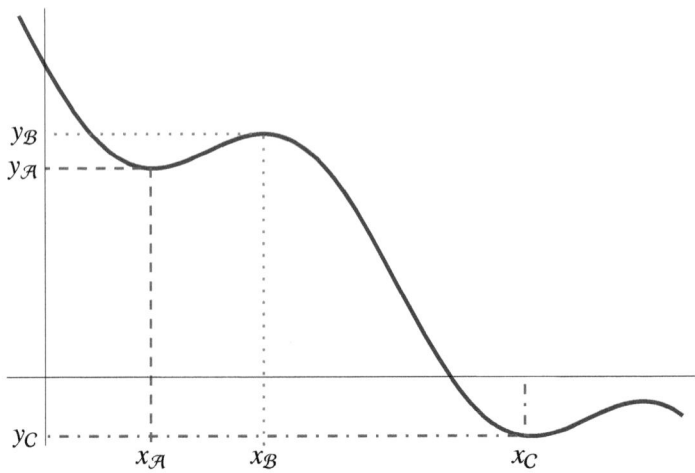

Figure 6.1: Exemplary smooth potential for typical escape problems.

now continue with an illustrative application of the quantum Smoluchowski equation (6.8), namely the escape from a metastable well.

6.1.4 Quantum enhanced escape rates

The problem of escape from a metastable well is a prototypical example in physics, which can be found e.g. in chemical reactions, diffusion in solids, or nuclear fission. Its particular feature is the separation of time scales between local well motion and long time decay characterized by a decay rate. The present subsection generalizes Kramers' original derivation [Kra40] to yield the semiclassical escape rate in the regime of the quantum Smoluchowski equation (6.8). In contrast to earlier publications [Ank01] the following considerations take care of the correct form of the quantum Smoluchowski equation (6.8) (see also [AG08]).

Let us start by considering a smooth potential, $V(x)$, with a metastable state as illustrated in Fig. 6.1. Moreover, we consider a semiclassical, Brownian particle to be initially trapped in the well at point $\mathscr{A} = (x_{\mathscr{A}}, y_{\mathscr{A}})$. Further, let the potential barrier, $\Delta = y_{\mathscr{B}} - y_{\mathscr{A}}$, be large compared to the thermal energy, $1/\beta$. We are interested in the rate for escapes from $\mathscr{A}$ to the next potential well at point $\mathscr{C} =$

6.1 Quantum Smoluchowski dynamics

$(x_\mathscr{C}, y_\mathscr{C})$. Initially the particle is prepared in its equilibrium distribution (6.11) in the well at $\mathscr{A}$. We assume the well at $\mathscr{C}$ to be another metastable well and, thus, we are merely interested in the rate for crossing over the single barrier at $\mathscr{B}$. In the stationary limit thermally activated escapes lead to a constant probability current,

$$j = -\frac{1}{\gamma M}\left(V'(x)p(x) + \frac{1}{\beta}\partial_x D_e(x)p(x)\right). \tag{6.12}$$

The latter current, j, in Eq. (6.12) can be rewritten in terms of the stationary distribution (6.11) as,

$$j = -\frac{p_s(x)D_e(x)}{\gamma M \beta}\partial_x \frac{p(x)}{p_s(x)}. \tag{6.13}$$

The diffusive current between the two metastable wells at $\mathscr{A}$ and $\mathscr{C}$ can, then, be obtained by integration and is given by,

$$j_{\mathscr{A}\to\mathscr{C}} = \frac{p(x_\mathscr{A})/p_s(x_\mathscr{A}) - p(x_\mathscr{C})/p_s(x_\mathscr{C})}{\gamma M \beta \int_\mathscr{A}^\mathscr{C} dx/p_s(x)D_e(x)}. \tag{6.14}$$

The latter equation enables us to derive an expression for the escape from a potential well over a barrier. Since we assume the system to be initially in a stationary state at $\mathscr{A}$, practically no particle has yet arrived at $\mathscr{C}$. Therefore, Eq. (6.14) can be simplified to read,

$$j_{\mathscr{A}\to\mathscr{C}} = \frac{p_\mathscr{A}}{\gamma M \beta \int_\mathscr{A}^\mathscr{C} dx/p_s(x)D_e(x)}, \tag{6.15}$$

where we introduced $p_\mathscr{A} = [p(x)/p_s(x)]_{\text{near }\mathscr{A}}$. The escape rate, $\Gamma_{\mathscr{A}\to\mathscr{C}}$, can, now, be defined as the ratio of the diffusive current, $j_{\mathscr{A}\to\mathscr{C}}$, and the number of particles, $n_\mathscr{A}$, in a small neighborhood around point $\mathscr{A}$,

$$\Gamma_{\mathscr{A}\to\mathscr{C}} = \frac{j_{\mathscr{A}\to\mathscr{C}}}{n_\mathscr{A}}. \tag{6.16}$$

The number of particles, $n_\mathscr{A}$, near $\mathscr{A}$ can be determined with the help of a harmonic approximation of the smooth potential, $V(x) \simeq \omega_\mathscr{A}^2 x^2/2$. Then $n_\mathscr{A}$ is given by,

$$n_\mathscr{A} = \int_{-\infty}^{+\infty} dx\, p_\mathscr{A} \exp\left(-\frac{\beta}{2}\omega_\mathscr{A}^2(1-\lambda\beta\omega_\mathscr{A}^2)x^2\right)\left(1-\lambda\beta\omega_\mathscr{A}^2\right). \tag{6.17}$$

6 Strong coupling limit - a semiclassical approach

The Gaussian integral in Eq. (6.17) can be evaluated and we obtain,

$$n_{\mathscr{A}} = p_{\mathscr{A}} \sqrt{\frac{2\pi\left(1-\lambda\beta\omega_{\mathscr{A}}^2\right)}{\beta\omega_{\mathscr{A}}^2}}. \tag{6.18}$$

Substituting the number of particles near $\mathscr{A}$, $n_{\mathscr{A}}$ in Eq. (6.18), in the escape rate, $\Gamma_{\mathscr{A}\to\mathscr{C}}$ in Eq. (6.16) we conclude,

$$\Gamma_{\mathscr{A}\to\mathscr{C}} = \frac{\omega_{\mathscr{A}}}{\gamma M \sqrt{2\pi\beta\left(1-\lambda\beta\omega_{\mathscr{A}}^2\right)}} \frac{1}{\int_{\mathscr{A}}^{\mathscr{C}} dx / p_s(x) D_e(x)}. \tag{6.19}$$

The main contribution to the integral is due to a small region near the barrier top at $\mathscr{B}$. Since we assume $V(x)$ to be smooth we can further approximate the potential in a small neighborhood around $\mathscr{B}$ as, $V(x) \simeq \Delta - \omega_{\mathscr{B}}^2(x-q)^2/2$, with $q = x_{\mathscr{B}} - x_{\mathscr{A}}$. Hence, the integral in Eq. (6.19) can be evaluated,

$$\int_{\mathscr{A}}^{\mathscr{C}} dx \frac{1}{p_s(x)D_e(x)} \simeq \exp(\beta\Delta) \int_{-\infty}^{+\infty} dx \exp\left(-\frac{\beta}{2}\omega_{\mathscr{B}}^2\left(1+\lambda\beta\omega_{\mathscr{B}}^2\right)(x-q)^2\right)$$

$$= \frac{\exp(\beta\Delta)\sqrt{2\pi}}{\sqrt{\beta\omega_{\mathscr{B}}^2\left(1+\lambda\beta\omega_{\mathscr{B}}^2\right)}}, \tag{6.20}$$

and, finally, the escape rate for quantum Smoluchowski dynamics results in,

$$\Gamma_{\mathscr{A}\to\mathscr{C}} \simeq \frac{\omega_{\mathscr{A}}\omega_{\mathscr{B}}}{2\pi\gamma M}\sqrt{\frac{\left(1+\lambda\beta\omega_{\mathscr{B}}^2\right)}{\left(1-\lambda\beta\omega_{\mathscr{A}}^2\right)}}\exp(-\beta\Delta). \tag{6.21}$$

The latter result for the escape rate, $\Gamma_{\mathscr{A}\to\mathscr{C}}$, reduces to the usual Kramers rate [Kra40] in the classical limit, $\lambda = 0$. Otherwise, for $\lambda \neq 0$ the rate is enhanced due to additional quantum fluctuations. The quantum fluctuations describe in a semiclassical framework the quantum peculiarity of tunneling through the potential barrier at $\mathscr{B}$. In Fig. 6.2 we plot the ratio of the quantum Smoluchowski rate, Γ_{QSE} (6.21), and the classical equivalent, Γ_{cl} (6.21) with $\lambda = 0$, for the simplest case of $\omega_{\mathscr{A}} \equiv \omega_{\mathscr{B}}$. We observe that the quantum fluctuations drastically increase the escape from a metastable well. Especially close to the boundary of the range of validity, $\lambda\beta\omega_{\mathscr{A}}^2 \lesssim 1$, the quantum effects govern the dynamics.

6.2 Quantum fluctuation theorems in the strong damping limit

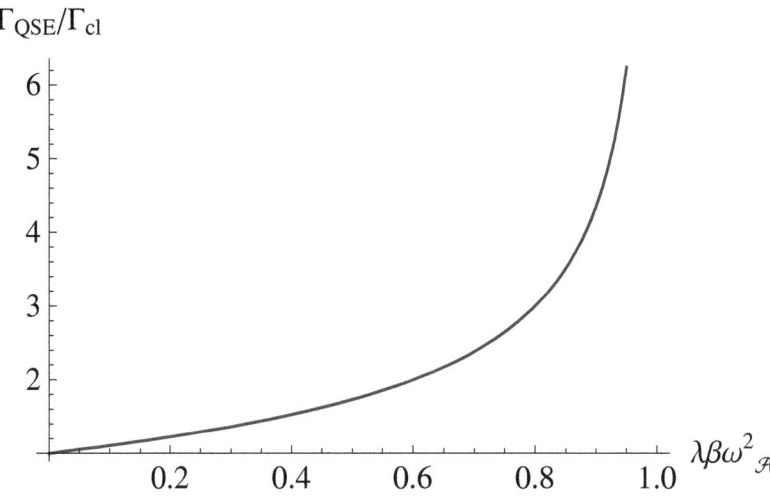

Figure 6.2: Ratio of quantum Smoluchowski and classical escape rate, Γ_{QSE}/Γ_{cl} (6.21), for the simple example $\omega_{\mathscr{A}} \equiv \omega_{\mathscr{B}}$.

6.2 Quantum fluctuation theorems in the strong damping limit

In the preceding section we presented the quantum Smoluchowski equation (6.8), which is a semiclassical model of an open quantum system strongly coupled to a thermal environment. In the present section (see also [BL11]) we, now, derive extensions of the Jarzynski and Crooks relations, Eqs. (2.36) and (2.57), by using a Wiener path integral representation of the solution of the quantum Smoluchowski equation following [CJ06]. For the sake of generality, we consider a generic driven Fokker-Planck equation, with position-dependent drift and diffusion coefficients, D_1 and D_2, of the form,

$$\partial_t p(x,\alpha,t) = \mathscr{F}_\alpha p(x,\alpha,t) , \qquad (6.22)$$

where the linear operator $\mathscr{F}_\alpha$ is given by,

$$\mathscr{F}_\alpha = -\partial_x D_1(x,\alpha) + \partial_x^2 D_2(x,\alpha) . \qquad (6.23)$$

6 Strong coupling limit - a semiclassical approach

The quantum Smoluchowski equation (6.8) corresponds to the particular choice $D_1(x,\alpha) = -V'(x,\alpha)/\gamma M$ and $D_2(x,\alpha) = 1/[1-\lambda\beta V''(x,\alpha)]\gamma M\beta$. In the present analysis, we consider a time-dependent problem where the potential $V(x,\alpha_t)$ is driven by some external parameter $\alpha_t = \alpha(t)$. The driving rate should be smaller than the relaxation rate, $\dot{\alpha}_t/\alpha_t \ll \gamma$, to ensure that the non-diagonal elements of the density operator remain negligibly small at all times [DL09]. Note that this condition is not restrictive in the limit of very large γ. Then, for any fixed value of the driving parameter, α, we write the stationary solution of Eq. (6.22) as,

$$p_s(x,\alpha) = \exp(-\varphi(x,\alpha)), \qquad (6.24)$$

where the function $\varphi(x,\alpha)$ is explicitly given by,

$$\varphi(x,\alpha) = \int^x dy \, \frac{\partial_y D_2(y,\alpha) - D_1(y,\alpha)}{D_2(y,\alpha)}. \qquad (6.25)$$

We denote by $X = \{x\}_{-\tau}^{+\tau}$ a stochastic trajectory of the system that starts at $t = -\tau$ and ends at $t = +\tau$. We further define a *forward process* α_t^F, in which the driving parameter is varied from an initial value, $\alpha_{-\tau}^F = \alpha_0$, to a final value, $\alpha_{+\tau}^F = \alpha_1$, as well as its time *reversed process*, $\alpha_t^R = \alpha_{-t}^F$. The conditional probability of observing a trajectory starting at $x_{-\tau}$ for the forward process can, then, be written as,

$$P^F[X|x_{-\tau}] = \exp\left(-\int_{-\tau}^{+\tau} dt \, S(x_t, \dot{x}_t, \alpha_t^F)\right), \qquad (6.26)$$

with a similar expression for the reversed process. For our purposes the stochastic action in Eq. (6.26), i.e. the generalized Onsager-Machlup functional $S(x_t, \dot{x}_t, \alpha_t)$, is taken to be of the form (cf. appendix C),

$$S(x_t, \dot{x}_t, \alpha_t) = \frac{[\dot{x}_t - (D_1(x_t, \alpha_t) - \partial_x D_2(x_t, \alpha_t))]^2}{4D_2(x_t, \alpha_t)}. \qquad (6.27)$$

The last term in the numerator of Eq. (6.27) is included to guarantee that thermodynamic potentials are independent of the state representation [GG79] and follows from the Itô-formula. By assuming that the system is initially in an equilibrium state given by the solution (6.24) of the Fokker-Planck equation (6.22), we obtain that the net probability of observing the trajectory X for the forward process is,

$$P^F[X] = p_s(x_{-\tau}, \alpha_0) P^F[X|x_{-\tau}]. \qquad (6.28)$$

6.2 Quantum fluctuation theorems in the strong damping limit

In complete analogy, we find that the corresponding unconditional probability for the reversed process reads,

$$P^R[X] = p_s(x_\tau, \alpha_1) P^R\left[X^\dagger | x_\tau\right], \quad (6.29)$$

where we introduced the time-reversed trajectory, $X^\dagger = \{x_t^\dagger\}_{-\tau}^{+\tau}$ with $x_t^\dagger = x_{-t}$. We next compare the probability of having the trajectory X during the forward process with that of having the trajectory $X^\dagger$ during the reversed process. We have

$$P^R\left[X^\dagger | x_{-\tau}^\dagger\right] = \exp\left(-\int_{-\tau}^{+\tau} dt\, S\left(x_t^\dagger, \dot{x}_t^\dagger, \alpha_t^R\right)\right) = \exp\left(-\int_{-\tau}^{+\tau} dt\, S^\dagger\left(x_t, \dot{x}_t, \alpha_t^F\right)\right), \quad (6.30)$$

where we defined the conjugate Onsager-Machlup function, as $S^\dagger(x_t, \dot{x}_t, \alpha_t) = S(x_t, -\dot{x}_t, \alpha_t)$. The ratio of the conditional probabilities (6.29) and (6.30) is simply determined by the difference of S and $S^\dagger$. Using the definition (6.27), we thus obtain,

$$\frac{P^F[X | x_{-\tau}]}{P^R\left[X^\dagger | x_{-\tau}^\dagger\right]} = \exp\left(\int_{-\tau}^{+\tau} dt\, \frac{D_1(x_t, \alpha_t^F) - \partial_x D_2(x_t, \alpha_t^F)}{D_2(x_t, \alpha_T^F)} \dot{x}_t\right). \quad (6.31)$$

The ratio of the forward and reversed probabilities, Eqs. (6.28) and (6.29), follows directly as,

$$\begin{aligned}\frac{P^F[X]}{P^R[X^\dagger]} &= \frac{p_s(x_{-\tau}, \alpha_0) P^F[X | x_{-\tau}]}{p_s(x_\tau, \alpha_1) P^R\left[X^\dagger | x_{-\tau}^\dagger\right]} \\ &= \exp\left(\Delta\varphi + \int_{-\tau}^{+\tau} dt\, \frac{D_1(x_t, \alpha_t^F) - \partial_x D_2(x_t, \alpha_t^F)}{D_2(x_t, \alpha_t^F)} \dot{x}_t\right),\end{aligned} \quad (6.32)$$

where $\Delta\varphi$ denotes the variation of the instantaneous stationary solution,

$$\Delta\varphi = \int_{-\tau}^{+\tau} dt\, \left(\dot{\alpha}_t^F \partial_\alpha \varphi + \dot{x}_t \partial_x \varphi\right). \quad (6.33)$$

6 Strong coupling limit - a semiclassical approach

By using the explicit expression (6.24) of the stationary solution $\varphi(x,\alpha)$, we finally arrive at

$$\frac{P^F[X]}{P^R[X^\dagger]} = \exp\left(\int_{-\tau}^{+\tau} dt\, \dot\alpha_t^F \,\partial_\alpha \varphi\right). \tag{6.34}$$

We are now in the position to derive generalized fluctuation theorems for stochastic processes described by the generic Fokker-Planck equation (6.22). We begin by defining the generalized entropy production Σ as,

$$\Sigma = \int_{-\tau}^{\tau} dt\, \dot\alpha_t^F \,\partial_\alpha \varphi. \tag{6.35}$$

The entropy production Σ in Eq. (6.35) is similar to the entropy production introduced by Hatano and Sasa for systems initially in a nonequilibrium steady state [HS01]. In the present situation, however, it corresponds to a non-Gibbsian equilibrium state. In addition, we note that the entropy production, as defined in Eq. (6.35), is odd under time-reversal, $\Sigma^R[X^\dagger] = -\Sigma^F[X]$. The distribution of the entropy production, $\rho^F(\Sigma)$, for an ensemble of realizations of forward processes can then be defined as,

$$\begin{aligned}\rho^F(\Sigma) &= \int \mathscr{D}X\, P^F[X]\, \delta\left(\Sigma - \Sigma^F[X]\right) \\ &= \exp(\Sigma) \int \mathscr{D}X^\dagger\, P^R\left[X^\dagger\right] \delta\left(\Sigma + \Sigma^R\left[X^\dagger\right]\right),\end{aligned} \tag{6.36}$$

where we used Eq. (6.34) in the last line. Moreover, the Wiener path integral,

$$\int \mathscr{D}X = \lim_{N\to\infty} (4\pi s)^{-N/2} \prod_{i=1}^{N-1} \int dx_{is} D(x_{is},\alpha_{is})^{-1/2} \tag{6.37}$$

with $s = 2\tau/N$, denotes the product of integrals over all possible, stochastic trajectories X. The continuous integral in Eq. (6.36) is interpreted as the limit of a discrete sum. Equation (6.34) can be recast in the form of a generalized Crooks relation for the entropy production,

$$\rho^R(-\Sigma) = \rho^F(\Sigma)\exp(-\Sigma). \tag{6.38}$$

By further integrating Eq. (6.38) over Σ, we obtain an extended version of the Jarzynski equality,

$$\langle \exp(-\Sigma) \rangle = 1 . \tag{6.39}$$

Expression (6.33) for the entropy production together with the fluctuation theorems (6.38) and (6.39) constitutes our main result of the present chapter. Combined, they represent the quantum generalizations of the Crooks and Jarzynski equalities, Eqs. (2.57) and (2.36), in the limit of strong damping. In the classical limit, $\lambda = 0$, $\varphi(x, \alpha) = \beta(V(x, \alpha) - F(\alpha))$, and the entropy production (6.35) takes the familiar form, $\Sigma = \beta \int dt\, \dot{\alpha}_t^F \partial_\alpha V(x_t, \alpha_t^F) - \beta \Delta F$. The inequality $\langle \Sigma \rangle \geq 0$ implied by Eq. (6.39) is often interpreted as an expression of the second law (cf. section 2.3). It is worthwhile to mention that the above derivation applies without modification to the case of an initial nonequilibrium steady state, instead of an initial equilibrium state, leading directly to a quantum generalization of the Hatano-Sasa relation [HS01]. Moreover, it is worth emphasizing that Σ (6.35) merely depends on the external driving, α_t, and the induced variation of the stationary solution, $\partial_\alpha \varphi$. Fluctuations of Σ solely stem from the explicit dependence of $\varphi(x_t, \alpha_t)$ on the stochastic position, x_t.

6.3 Experimental verification in Josephson junctions

No experimental investigation of any quantum fluctuation theorem has been performed so far. A scheme to study the Crooks and Jarzynski relations in isolated and weakly damped quantum systems using modulated ion traps has recently been put forward in [DL08a] (cf. section 4.5). In the present section we turn to a possible experimental verification of the above derived fluctuation theorems (6.38) and (6.39) in the strong coupling regime. The prototypical, physical system described by the quantum Smoluchowski equation (6.8) consists of an overdamped Josephson junction [Ank04, TC08, TC09, uH06]. A Josephson junction is built by two superconductors separated by a thin barrier through which Cooper pairs can tunnel. In the following, we will discuss how the dynamics of Josephson junctions can be described mathematically by means of the quantum Smoluchowski approach. Moreover, we will present the resulting I-V characteristics including quantum fluctuations and, finally, propose a possible measurement procedure for

131

6 Strong coupling limit - a semiclassical approach

the verification of the generalized expressions of the second law.

6.3.1 RCSJ-model

In order to describe the dynamics of Josephson junctions one has to consider situations where the junction current is larger than the maximum Josephson current, I_c. For such set-ups an externally applied current can not be completely carried by the Josephson current, I_s. Hence, we have to include additional channels carrying the excess current in the description. For temperatures larger than zero, there is a finite probability for the Cooper pairs to break up into *quasiparticles*. These quasiparticles contribute similar to *normal* electrons *resistively* to the current by a present finite voltage over the junction. The second additional channel is given by the finite *capacitance* of real Josephson junctions. An equivalent circuit is plotted in Fig. 6.3 to illustrate the situation. Due to Kirchhoff's law [Lik86] the

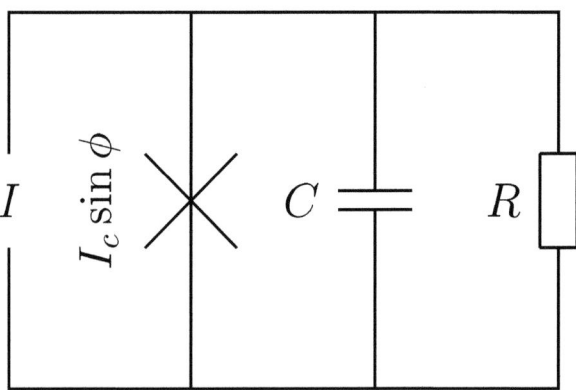

Figure 6.3: Equivalent circuit for a resistively and capacitively shunted Josephson junction (RCSJ-model)

net current, I, flowing through the Josephson junction can be separated into four contributions,

$$I = I_s + I_R + I_C - \mathfrak{J}_t, \tag{6.40}$$

6.3 Experimental verification in Josephson junctions

where I_R is the resistive and I_C the capacitive current. Moreover, we include a fluctuating term, $\mathfrak{J}_t$, stemming e.g. from the Nyquist noise in electrical circuits. The resistive current is carried by the quasiparticles. Therefore, we identify I_R with the help of the usual Ohm's law,

$$I_R = \frac{U}{R}, \tag{6.41}$$

where U is the voltage drop across the junction and R the resistance. For the dynamic case, where not only U_t, but also its time derivative, $d_t U_t$, is nonzero, the capacitive current reads,

$$I_C = C\,d_t U_t, \tag{6.42}$$

with the capacitance C. Finally, the Josephson current, I_s, is determined by the Josephson equations [Jos62],

$$I_s = I_c \sin(\phi_t) \quad \text{and} \quad d_t \phi_t = \frac{2e}{\hbar} U_t. \tag{6.43}$$

The maximal current, I_c, is given by $I_c = (2e/\hbar) E_J$, where E_J is the coupling energy (Josephson energy). The phase difference between left and right superconductor is denoted by ϕ. Substituting Eqs. (6.41)-(6.43) in the net current (6.40) we obtain,

$$\begin{aligned} I_t &= I_c \sin(\phi_t) + \frac{U_t}{R} + C\,d_t U - \mathfrak{J}_t \\ &= I_c \sin\phi_t + \frac{\hbar}{2e}\frac{1}{R} d_t \phi_t + \frac{\hbar}{2e} C\,d_t^2 \phi_t - \mathfrak{J}_t, \end{aligned} \tag{6.44}$$

where $\Phi_0 = h/2e$ is the magnetic flux quantum. The latter equation can be rewritten as,

$$\left(\frac{\hbar}{2e}\right)^2 C\,d_t^2 \phi_t + \left(\frac{\hbar}{2e}\right)^2 \frac{1}{R} d_t \phi_t + \frac{\hbar}{2e} \partial_\phi \left(-I_c \cos(\phi_t) - I_t \phi_t\right) = \mathfrak{J}_t. \tag{6.45}$$

Hence, we conclude that the dynamics of Josephson junctions can be interpreted as the diffusive motion of a particle with position ϕ_t and mass $M = (\hbar/2e)^2 C$, and the friction coefficient is given by $\gamma = 1/RC$. Moreover, the external potential is identified as,

$$V(\phi,t) = -E_J \cos(\phi) - E_I(t)\,\phi, \tag{6.46}$$

6 Strong coupling limit - a semiclassical approach

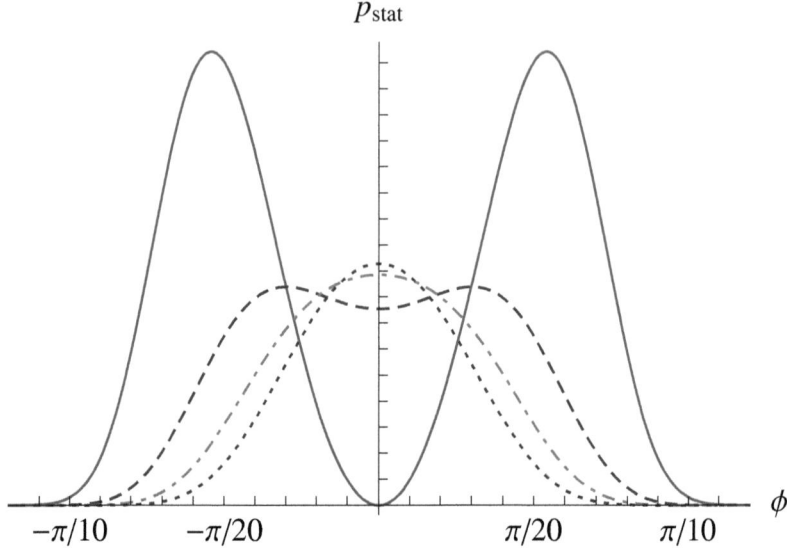

Figure 6.4: Stationary solution (6.49) for a circle-shaped junction with $C = 1.2pF$, $R = 0.37\Omega$, $I_c = 0.2mA$ and $\beta_c = 0.1$ and different temperature with: $\theta \simeq 1$ and $T = 0.973K$ (solid line); $\theta = 0.99$ and $T = 0.979K$ (large-dashed line); $\theta = 0.98$ and $T = 0.986K$ (dot-dashed line); $\theta = 0.97$ and $T = 0.993K$ (dashed line).

where the energy, $E_I(t) = (\hbar/2e)I_t$, is determined by the bias current, I_t. An important quantity in the RCSJ model is the dimensionless capacitance (Stewart-McCumber parameter), $\beta_c = (2\pi/\Phi_0)I_c R^2 C$, where $\Phi_0 = h/2e$ is the magnetic flux quantum. The inertial term in Eq. (6.45) becomes negligible for $\beta_c \ll 1$ and, thus, the magnitude of the Stewart-McCumber parameter determines the overdamped regime. Furthermore, it is worth emphasizing that the fluctuating current, $\mathfrak{J}_t$, is completely determined by the thermal environment surrounding the Josephson junction. Hence, for small temperatures the Josephson dynamics are governed by thermal as well as by quantum fluctuations. In the overdamped case the quantum Smoluchowski equation becomes applicable and Eq. (6.8) equivalently describes the dynamics (cf. subsection 2.2.2). The effective diffusion coefficient

6.3 Experimental verification in Josephson junctions

(6.9) reads,

$$D_e(\phi) = \frac{1}{1 - \theta \cos(\phi)}. \tag{6.47}$$

The constant $\theta = \lambda \beta E_J$ is the crucial parameter which governs the magnitude of quantum effects in a Josephson junction. It is directly proportional to the quantum parameter λ in Eq. (6.10) which in the context of the RCSJ model can be reexpressed as,

$$\lambda = 2\rho \left[c + \Psi \left(\frac{\beta E_c}{2\pi^2 \rho} + 1 \right) \right], \tag{6.48}$$

where $E_c = 2e^2/C$ is the charging energy, $\rho = R/R_Q$ the dimensionless resistance and $R_Q = h/4e^2$ the resistance quantum. The stationary solution of the quantum Smoluchowski equation (6.3) with periodic boundary conditions, i.e. $p_{\text{stat}}(\phi) = p_{\text{stat}}(\phi + L)$, can be written as [Rei02],

$$p_{\text{stat}}(\phi) = \frac{p_s(\phi)}{Z_J} \int_\phi^{\phi+L} \frac{dy}{D_e(y) p_s(y)}, \tag{6.49}$$

where $p_s(\phi)$ is given by Eq. (6.11) and Z_J is the normalization constant. In Fig. 6.4 we plot the stationary solution (6.49) for realistic experimental values of a niobium-based junction [KS10]. We observe that with decreasing temperature, i.e. increasing values of θ, the quantum fluctuations drive the semiclassical particle from the potential well. Moreover, the quantum effects depend sensitively on the temperature. For higher temperatures the distributions show the classically expected shape with their maximum at the potential minimum, $\phi = 0$. By further cooling down the probability to find a particle at the minimum of the potential withers, since the local diffusion is proportional to the curvature. Hence, for very small temperatures the overdamped quantum particle is most likely found outside the potential well. Nevertheless, by varying the temperature, both the classical, $\theta \ll 1$, as well as the quantum regime, $\theta \lesssim 1$, can be explored with the same junctions.

6.3.2 I-V characteristics

In a first step for the verification of the fluctuation theorems (6.38) and (6.39) the applicability of the quantum Smoluchowski description has to be confirmed. Experimentally, the I-V curve is a significant characteristic. Hence, the present

6 Strong coupling limit - a semiclassical approach

subsection is dedicated to the generalization of the result of Ambegaokar and Halperin [AH69] for a Josephson junction in the quantum Smoluchowski regime. The present derivation follows the main steps of [AH69]. To this end, we review and simplify the original derivation and propose the quantum generalization. For the sake of clarity we introduce the parameter $\mathfrak{a}$ and the dimensionless current $\mathfrak{i}$,

$$\mathfrak{a} = \frac{\hbar \beta I_c}{e} \quad \text{and} \quad \mathfrak{i} = \frac{I}{I_c}. \tag{6.50}$$

Hence, the potential, $V(\phi)$ in Eq. (6.46), can be rewritten as,

$$V(\phi) = -\frac{\mathfrak{a}}{2\beta}(\cos(\phi) + \mathfrak{i}\phi). \tag{6.51}$$

Now, we express the quantum Smoluchowski equation (6.8) in terms of the probability current, j, with the help of a particular solution, $\mathfrak{s}(\phi)$,

$$\frac{1}{\gamma M} \partial_\phi \left[V'(\phi) + \frac{1}{\beta} \partial_\phi D_e(\phi) \right] \mathfrak{s}(\phi) \equiv -\partial_\phi j. \tag{6.52}$$

For the derivation of the I-V curve we are interested in the stationary state, where j is constant. As noted earlier, for periodic boundary conditions as implied by potentials like in Eq. (6.51) we only need to consider the value of ϕ modulo 2π (cf. Eq. (6.49)). Consequently, we may restrict ϕ to the interval $0 \leq \phi \leq 2\pi$. For normalized stationary solutions, $\mathfrak{s}_{\text{stat}}(\phi)$, the inverse probability current, $1/j$, is given by the mean first passage time, i.e. the average time for a phase slippage of 2π. Thus, the mean voltage $\langle U \rangle$ is given in accordance with the Josephson equations (6.43) by,

$$\langle U \rangle = \frac{h}{2e} j. \tag{6.53}$$

For the present purpose the stationary solution (6.49) is reformulated with $L = 2\pi$ to read,

$$\mathfrak{s}_{\text{stat}}(\phi) = j M \gamma \beta \frac{p_s(\phi)}{p_s(2\pi) - p_s(0)} \\
\times \left[p_s(0) \int_0^\phi \frac{d\eta}{D_e(\eta) p_s(\eta)} + p_s(2\pi) \int_\phi^{2\pi} \frac{d\eta}{D_e(\eta) p_s(\eta)} \right], \tag{6.54}$$

6.3 Experimental verification in Josephson junctions

where $p_s(\phi)$ is again the stationary solution (6.11) normalized over the total support of the quantum Smoluchowski equation (6.8). From the latter formulation of the periodic stationary solution (6.54) and noting its normalization over one period, $\int_0^{2\pi} d\phi\, \mathfrak{s}_{\text{stat}}(\phi) = 1$, the dimensionless voltage $v = \langle U \rangle / I_c R$ is obtained to read,

$$v = \frac{4\pi}{a} \frac{\exp(a\pi i) - 1}{\exp(a\pi i)} \left[\int_0^{2\pi} d\phi \int_0^{2\pi} d\eta \, \frac{p_s(\phi)}{D_e(\eta) p_s(\eta)} \right]^{-1}. \quad (6.55)$$

The latter equation constitutes the generalization of the classical I-V characteristics to systems describable by means of the quantum Smoluchowski equation (6.8). In the following we continue with a graphical analysis of the analytical I-V characteristics (6.55). To this end, the double integral expression in Eq. (6.55)

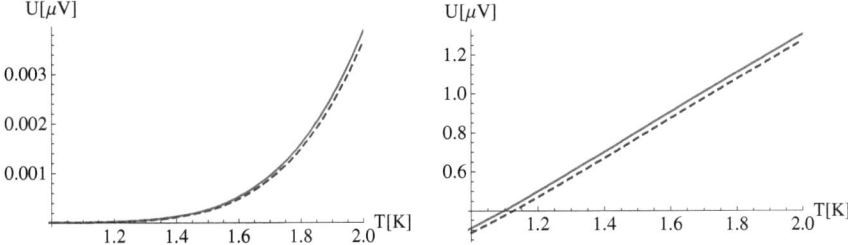

Figure 6.5: Temperature dependent I-V characteristics (solid line) (6.55) together with classically expected results (dashed line) (6.60) for circle shaped Josephson junctions with $C = 1.2 pF$, $R = 0.37\Omega$, $I_c = 0.2 mA$ and $\beta_c = 0.1$ and different values of i; left, i = 0.985; right, i = 0.995;

has to be evaluated numerically. For the sake of higher numerical accuracy and comparability with the classically expected result [AH69] we further reformulate the I-V curve (6.55) with the help of a coordinate transformation. Let Υ be a two dimensional map,

$$\begin{aligned} \Upsilon : \quad [\tfrac{\pi}{2}, \tfrac{5}{2}\pi] \times [0, 2\pi] &\to [0, 2\pi]^2 \\ (x, y) &\mapsto (\phi, \eta) \end{aligned} \quad (6.56)$$

6 Strong coupling limit - a semiclassical approach

whose inverse map transforms the original variables, (ϕ, η), to the new variables, (x, y), and explicitly reads,

$$x = \frac{1}{2}(\eta + \phi + \pi) \quad \text{and} \quad y = \eta - \phi \qquad (6.57a)$$

and

$$\phi = \frac{1}{2}(2x - y - \pi) \quad \text{and} \quad \eta = \frac{1}{2}(2x + y - \pi). \qquad (6.57b)$$

Then, the I-V characteristics (6.55) can be reformulated as a two dimensional integral expression in the new variables, (x, y),

$$v = \frac{4\pi}{a} \frac{\exp(a\pi i) - 1}{\exp(a\pi i)}$$

$$\times \left\{ \int_0^{2\pi} dy \int_{\pi/2}^{5/2\pi} dx \left[1 - \frac{\lambda a}{2} \left(\sin(x) \cos\left(\frac{y}{2}\right) - \sin\left(\frac{y}{2}\right) \cos(x) \right) \right] \right.$$

$$\times \exp\left(-ai\frac{y}{2} - a\sin\left(\frac{y}{2}\right) \cos(x)\right) \qquad (6.58)$$

$$\left. \times \exp\left(\frac{\lambda a^2}{8} \left[\sin(2x) \sin(y) + 4i \sin(x) \sin\left(\frac{y}{2}\right) \right] \right) \right\}^{-1}.$$

The latter expression in Eq. (6.58) further simplifies in the classical limit, $\lambda = 0$, and we have,

$$v_{cl} = \frac{4\pi}{a} \frac{\exp(a\pi i) - 1}{\exp(a\pi i)} \left[\int_0^{2\pi} dy \int_{\pi/2}^{5/2\pi} dx \exp\left(-ai\frac{y}{2}\right) \exp\left(-a\sin\left(\frac{y}{2}\right) \cos(x)\right) \right]^{-1}, \qquad (6.59)$$

which can be simplified with the help of the modified Bessel function I_0 [AS72] to read,

$$v_{cl} = \frac{2}{a} \frac{\exp(a\pi i) - 1}{\exp(a\pi i)} \left[\int_0^{2\pi} dy \exp\left(-ai\frac{y}{2}\right) I_0\left(a\sin\left(\frac{y}{2}\right)\right) \right]^{-1}. \qquad (6.60)$$

6.3 Experimental verification in Josephson junctions

Equation (6.60) is the I-V curve, which is expected to be measured if quantum effects of a low temperature bath are not taken into account. Moreover, Eq. (6.60) expresses the result of [AH69] and coincides with the formula given in the lecture notes of Gross and Marx [GM05].

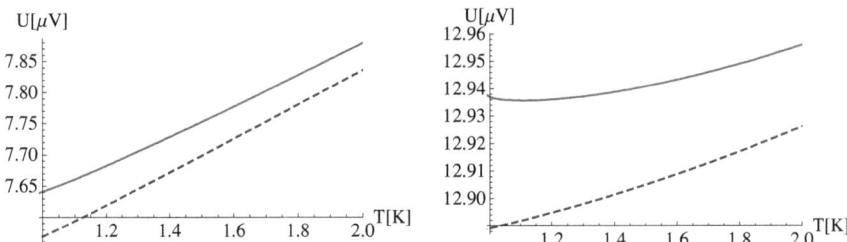

Figure 6.6: Temperature dependent I-V characteristics (solid line) (6.55) together with classically expected results (dashed line) (6.60) for circle shaped Josephson junctions with $C = 1.2pF$, $R = 0.37\Omega$, $I_c = 0.2mA$ and $\beta_c = 0.1$ and different values of i; left, $i = 1.005$; right, $i = 1.015$;

In Figs. 6.5 and 6.6 we plot the quantum Smoluchowski I-V characteristics (6.55) together with the classically expected result (6.60) for realistic Josephson junctions [KS10]. For illustration of the quantum effects the voltage, U, is plotted as a function of the temperature for different values of the dimensionless bias current, i. We observe that the quantum fluctuations lead to an enhanced voltage, U, and, thus, manifest themselves in an apparently higher temperature in the measurement outcome. The lower temperature range in the plots in Figs. 6.5 and 6.6 is determined by the conditions of validity of the quantum Smoluchowski equation (6.6). For even lower temperatures a complete quantum mechanical description becomes necessary [IP98, IG99]. Finally we note that the additional quantum fluctuations imply for low temperatures a significant effect on the I-V characteristics. However, highly sensitive measurements become necessary to exclude measurement errors on small voltage scales. Moreover, the quantum effects in the I-V curves are largest for bias currents $i \gtrsim 1$ (cf. Fig. 6.6) and are negligible for $i \lesssim 1$ (cf. Fig. 6.6).

6 *Strong coupling limit - a semiclassical approach*

6.3.3 Possible measurement procedure

Once it has been shown that the quantum Smoluchowski description is applicable, the fluctuation theorems (6.38) and (6.39) can be experimentally verified in such systems. The nonequilibrium entropy production (6.35) can be experimentally determined in a Josephson junction by applying the following measurement procedure. The phase ϕ can be directly deduced from a measurement of the Josephson current, I_s, once the current-phase relation of the junction has been determined [KI04]. The system is then first prepared in a given initial state and let to relax to its stationary state (6.49). After the latter has been attained, the Josephson potential, $V(\phi)$, is modified according to a specific driving protocol, α_t, with e.g the help of an external magnetic field. The entropy production, Σ, during such a protocol (corresponding to either a *forward* or *reversed* transformation) can be evaluated via Eqs. (6.35) and (6.43) from the recorded values of the current. The distribution function of the entropy can eventually be reconstructed by repeating the above measurement sequence, and the validity of the quantum fluctuation theorems (6.38) and (6.39) in the quantum Smoluchowski regime can be tested. For a Josephson junction, which is driven by a time-dependent bias current, the entropy distributions have been determined and the fluctuation theorems (6.38) and (6.39) verified numerically in [Bru10]. The driving with a time-dependent bias current has the advantage that the complexity of the experimental set-up is reduced by omitting additional wires inside the cryostat generating the predetermined magnetic field. However, the external driving, i.e. here outside the cryostat, is more sensitive to external perturbations. The experimental realization of junctions with the appropriate parameters has already been possible, whereas the details of the *real* measurement procedure still have to be clarified [KS10].

6.4 Summary

In the present chapter we considered an open quantum system strongly coupled to a thermal environment. For such quantum systems a semiclassical description becomes feasible due to generic effects of the environment, like decoherence. First, we presented the quantum Smoluchowski equation, which is a classical equation of motion with a quantum modified diffusion coefficient. In a semiclassical picture quantum effects manifest themselves as additional quantum fluctuations and,

6.4 Summary

hence, an effective diffusion coefficient. As an illustration of the properties we derived the escape rate from a metastable well and showed that the quantum fluctuations properly describe quantum peculiarities like tunneling through a potential barrier. As a completely new contribution we, then, derived Crooks and Jarzynski type fluctuation theorems for the reduced system dynamics. The latter expressions of the second law for quantum systems arbitrarily far from thermal equilibrium can be verified in a Josephson junction experiment. To this end, we introduced the RCSJ-model and discussed, how the Josephson junctions dynamics are translated into the quantum Smoluchowski regime. As an experimental verification of the applicability of the model we proposed the measurements of the I-V characteristics. We concluded that the quantum effects lead to an enhanced voltage in the I-V curve for particular values of the temperature and the bias current as it is expected from the classical theory. Finally, we proposed a possible measurement procedure for the verification of the fluctuation theorems.

6 Strong coupling limit - a semiclassical approach

7 Epilogue

The present thesis covered a wide range of problems in quantum thermodynamics. Starting with a geometric approach to explain the dynamical properties of isolated quantum systems in chapter 3 we developed a complete thermodynamic description of quantum systems obeying unitary dynamics arbitrarily far from equilibrium in chapter 4. As a prototypic experimental system for the research of quantum thermodynamics we analyzed cold ion traps. Then, we turned to quantum systems coupled to a thermal environment. In chapter 5 we derived the universal expression for the irreversible entropy production and an integral fluctuation theorem valid in the weak coupling regime. Moreover, we were able to derive the equivalent result in the strong damping limit in chapter 6, which could be verified with the help of a Josephson junction experiment. Nevertheless, there remain various open problems, which have to be solved in future research. In particular, the development of a general approach to thermodynamics considering merely the reduced system dynamics is still lacking. Admittedly, the present thesis and recent publications [TH09a, TH09b, CH09] cover most physically relevant situations. However, from an experimental point of view, usually, only the system of interest is controllable. On the contrary, the heat bath may be arbitrarily large, and, thus, it is not possible to gain complete information. Moreover, we only have resilient results in the weak and high damping limit. For general couplings of the reduced system to arbitrary environments a completely thermodynamic description is an unsolved problem.

One might pose the question, whether it makes sense to insist on a thermodynamics description of the reduced system only. Over the last two centuries thermodynamics has been proven to be an extremely powerful concept, since it is a phenomenological theory of mean values. From chemical reactions to black holes thermodynamics can explain the physics of general systems at any length scale. For systems arbitrarily far from equilibrium, however, the physical properties are significantly dependent on the explicit time evolution. For classical systems as well as for quantum systems ultra-weakly coupled to a thermal environment it has been possible to extend the conventional thermodynamic theory explicitly tak-

7 Epilogue

ing fluctuations into account. On the contrary, for open quantum systems with arbitrary coupling to the environment dynamics are governed by the interaction between the distinct subsystems. Hence, thermodynamic quantities like work and heat have to be re-defined carefully. Thus, it is desirable to develop the corresponding theory.

Concluding remarks

The present thesis drew a bow over a wide range of couplings between a quantum system of interest and a thermal environment. We have been able to derive general expressions for the irreversible entropy production for any kind of processes operating arbitrarily far from thermal equilibrium. Starting with a phenomenological approach to the accessible degrees of freedom we developed general expressions of the second law for almost all imaginable, physically relevant quantum situations. The present considerations have been motivated by an experimental point of view and the original philosophy of thermodynamics. To this end, we had to elaborate a geometric approach to isolated quantum systems and include methods of statistical physics, quantum information theory and the theory of open quantum systems.

A Quantum information theory

The present appendix introduces information theoretic quantities, which have been used in the above considerations. We summarize definitions and outline the most important properties.

A.1 Relative entropy

The Shannon or von Neumann entropy [vN55] is a function that is an expression of the information content of a probability distribution. However, in information theory one often wants to measure the distinguishability of two distributions, $p(x)$ and $q(x)$. One of the most frequently considered quantities is the Kullback-Leibler divergence [KL51],

$$D(p||q) = \int dx\, p(x) \ln\left(\frac{p(x)}{q(x)}\right). \tag{A.1}$$

The latter equation defines a non-symmetric measure of the difference between two probability distributions. Moreover, $D(p||q)$ measures the expected number of extra bits required to code samples from $p(x)$ when using a code based on $q(x)$, rather than using a code based on $p(x)$. The divergence is always non-negative,

$$D(p||q) \geq 0, \tag{A.2}$$

which is a result known as Gibbs' inequality, with $D(p||q)$ equals zero only for identical distributions, $p \equiv q$. Furthermore, the divergence remains well-defined for all continuous distributions and is invariant under parameter transformations.

A Quantum information theory

A.1.1 Inequalities in information theory

The most important inequalities in information theory are lower bounds on the Kullback-Leibler divergence (A.1). Even the Shannon-type inequalities can be recast in this category. There exist only very rare results for useful upper bounds, since the divergence $D(p||q)$ depends very sensitively on events that are very rare in the reference distribution q. Hence, $D(p||q)$ diverges if an event of finite non-zero probability in p becomes exceedingly rare in q. Therefore, p is required to be absolutely continuous with respect to q to avoid $D(p||q)$ being ill-defined.

Gibbs' inequality

The Gibbs' inequality was first proposed in the 19th century. For discrete probability distributions, $p_n \in \{p_1, ..., p_N\}$ and $q_n \in \{q_1, ..., p_N\}$, the following inequality between positive quantities holds,

$$-\sum_{n=1}^{N} p_n \ln(p_n) \leq -\sum_{n=1}^{N} q_n \ln(p_n), \tag{A.3}$$

with equality only if $p_n = q_n$ for all n. A direct corollary is the non-negativity of Kullback-Leibler divergence, $D(p||q)$, in Eq. (A.2).

Kullback's inequality

The Kullback's inequality gives a lower bound on $D(p||q)$ expressed in terms of the large deviations rate function. It reads,

$$D(p||q) \geq \Psi_q^*(\mu(p)), \tag{A.4}$$

where Ψ_q^* is the rate function, i.e. the convex conjugate of the cumulant-generating function, of q, and $\mu(p)$ is the first moment of p. A corollary of the latter Eq. (A.4) is the Cramér-Rao bound [Rao45] (see also Eq. (A.19)).

A.1.2 Quantum relative entropy

In quantum information theory, the quantum relative entropy is the quantum mechanical generalization of the Kullback-Leibler divergence (A.1). It was first pro-

A.2 Fisher information

posed by Umegaki [Ume62] and reads for two density operators, ρ_1 and ρ_2,

$$S(\rho_1||\rho_2) = \text{tr}\{\rho_1 \ln \rho_1\} - \text{tr}\{\rho_1 \ln \rho_2\}. \tag{A.5}$$

We immediately see from Eq. (A.5) that for commuting densities, $[\rho_1, \rho_2] = 0$, the definition (A.5) coincides with the classical case (A.1). Further, the quantum relative entropy is always non-negative, $S(\rho_1||\rho_2) \geq 0$, which is usually known as Klein's inequality. Finally, $S(\rho_1||\rho_2)$ serves as a measure of entanglement [Ved02]. Let a composite quantum system have a Hilbert space, $\mathcal{H}$, with

$$\mathcal{H} = \otimes_k \mathcal{H}_k, \tag{A.6}$$

and ρ is a density operator acting on the total system. Then, the relative entropy of entanglement of ρ is defined as,

$$D_{\text{ent}}(\rho) = \min_{\sigma} \{S(\rho||\sigma)\}, \tag{A.7}$$

where the minimum is taken over all separable states, σ. It follows, that ρ is not an entangled state, if $D_{\text{ent}}(\rho) = 0$.

A.2 Fisher information

The Fisher information, $\mathscr{I}(\Theta)$, is the variance of the score. This means, that $\mathscr{I}(\Theta)$ measures the amount of information that an observable, random variable, x, carries about an unknown parameter, Θ. We denote the likelihood function of Θ by $p(x;\Theta)$. The likelihood function is the joint probability of the data, the values of x, conditional on the value of Θ. The variance of the score is simply given by the second moment of the score, since the expectation is zero. Hence, the Fisher information reads,

$$\mathscr{I}(\Theta) = \left\langle [\partial_\Theta \ln(p(x;\Theta))]^2 \right\rangle_\Theta. \tag{A.8}$$

For parametric probability densities, $p(x;\Theta)$, Eq. (A.8) can be written as,

$$\mathscr{I}(\Theta) = \int dx \frac{(\partial_\Theta p(x;\Theta))^2}{p(x;\Theta)}. \tag{A.9}$$

A Quantum information theory

From the latter formulation in Eq. (A.9) of $\mathscr{I}(\Theta)$ we observe that the infinitesimal Fisher information is given by the square of the infinitesimal statistical distance (3.11),

$$d\mathscr{I} = \left(\frac{1}{2}d\ell\right)^2. \tag{A.10}$$

The quantum generalization of the latter Eq. (A.10) was used in Eq. (3.58).

A.2.1 Relation to Kullback-Leibler divergence

Now, we consider two parametric probability distributions, $p(\Theta)$, with $p_1 = p(\Theta_1)$ and $p_2 = p(\Theta_2)$. Then, the Kullback-Leibler divergence (A.1) reads,

$$D(p(\Theta_1) || p(\Theta_2)) = \int dx\, p(x; \Theta_1) \ln\left(\frac{p(x; \Theta_1)}{p(x; \Theta_2)}\right). \tag{A.11}$$

For infinitesimal changes in the parameter Θ, $\Theta_2 = \Theta_1 + \delta\Theta$, the divergence is proportional to the Fisher information (A.8) [Kul78],

$$D(p(\Theta_2 - \delta\Theta) || p(\Theta_2)) \xrightarrow{\delta\Theta \to 0} \frac{1}{2}\mathscr{I}(\Theta_2)(\delta\Theta)^2, \tag{A.12}$$

why we conclude,

$$\partial_\Theta^2 D(p(\Theta) || p(\Theta_2))|_{\Theta=\Theta_2} = \mathscr{I}(\Theta_2). \tag{A.13}$$

The latter equation is the underlying quantum information theoretic relation which led to the generalization of the Clausius inequality in chapter 4.

A.2.2 Cramér-Rao bound

The Cramér-Rao bound relates the Fisher information with the quality of a measurement with estimators $\hat{\Theta}(x)$. We consider in the following the class of unbiased estimates, obeying $\langle\hat{\Theta}(x)\rangle = \Theta$. Here, the angle brackets denote the average over the probability density, $p(x; \Theta)$,

$$\langle\hat{\Theta}(x) - \Theta\rangle_\Theta = \int dx\, (\hat{\Theta}(x) - \Theta)\, p(x; \Theta) = 0. \tag{A.14}$$

By differentiating with respect to Θ, the latter equation transforms into,

$$\int dx \, (\hat{\Theta}(x) - \Theta) \, \partial_\Theta p(x;\Theta) - \int dx \, p(x;\Theta) = 0. \tag{A.15}$$

With the identity $\partial_\Theta p = p \partial_\Theta \ln(p)$ and making use of the normalization of p Eq. (A.15) reads,

$$\int dx \, (\hat{\Theta}(x) - \Theta) \, p(x;\Theta) \, \partial_\Theta \ln(p(x;\Theta)) = 1, \tag{A.16}$$

which we rewrite as,

$$\int dx \, [\sqrt{p} \, \partial_\Theta \ln(p(x;\Theta))] \left[(\hat{\Theta}(x) - \Theta) \sqrt{p} \right] = 1. \tag{A.17}$$

By squaring Eq. (A.17) and making use of the Cauchy-Schwartz inequality we, finally, obtain,

$$\left\langle [\partial_\Theta \ln(p(x;\Theta))]^2 \right\rangle_\Theta \left\langle [\hat{\Theta}(x) - \Theta]^2 \right\rangle_\Theta \geq 1, \tag{A.18}$$

which is the Cramér-Rao bound [Rao45]. Hence, the Fisher information, $\mathscr{I}(\Theta)$, estimates the precision of a measurement from below,

$$\left\langle [\hat{\Theta}(x) - \Theta]^2 \right\rangle_\Theta \geq \frac{1}{\mathscr{I}(\Theta)}. \tag{A.19}$$

It is worth emphasizing that the Fisher information, $\mathscr{I}(\Theta)$, is a fundamental quantity in information theory, whose relation to thermodynamics is not completely clarified yet.

A.3 Bures metric

In the above considerations we introduced the Bures length (3.30) and the Bures distance (4.33). The underlying metric of both quantities first appeared in the literature by Kakutani in 1948 [Kak48] and was mathematically properly defined by Bures in 1969 [Bur69]. The Bures metric defines the infinitesimal distance between density operators,

$$[\mathscr{L}(\rho, \rho + \delta\rho)]^2 = \frac{1}{2} \text{tr}\{\delta\rho \, G\}, \tag{A.20}$$

A Quantum information theory

where G is implicitly given by,

$$\rho G + G\rho = \delta\rho. \tag{A.21}$$

Therefore, the Bures metric is by definition a non-trivial quantity, whose explicit expression can only be evaluated for special situations.

A.3.1 Explicit formulas

From the definition (A.20) the evaluation of the Bures metric is not evident. For finite dimensional systems Dittmann [Dit99] proposed the following formulas valid for 2×2 and 3×3 systems, respectively,

$$[\mathscr{L}(\rho, \rho + \delta\rho)]^2 = \frac{1}{4} \mathrm{tr} \left\{ \delta\rho^2 + \frac{1}{\det\rho} (1-\rho)\delta\rho(1-\rho)\delta\rho \right\}, \tag{A.22}$$

and

$$\begin{aligned}
[\mathscr{L}(\rho, \rho + \delta\rho)]^2 &= \frac{1}{4} \mathrm{tr} \left\{ \delta\rho^2 + \frac{3}{1-\mathrm{tr}\{\rho^3\}} (1-\rho)\delta\rho(1-\rho)\delta\rho \right\} \\
&+ \frac{1}{4} \mathrm{tr} \left\{ \frac{3\det\rho}{1-\mathrm{tr}\{\rho^3\}} \left(1-\rho^{-1}\right)\delta\rho\left(1-\rho^{-1}\right)\delta\rho \right\}.
\end{aligned} \tag{A.23}$$

Moreover, in terms of the eigenvectors $|n\rangle$ and eigenvalues p_n of the density operator ρ Hübner found [Hue92],

$$[\mathscr{L}(\rho, \rho + \delta\rho)]^2 = \frac{1}{2} \sum_{i,j=1}^n \frac{|\langle j|\delta\rho|k\rangle|^2}{p_j + p_k}. \tag{A.24}$$

Finally, following Sommers and Życzkowski [SZ03] the volume element induced by the Bures metric (A.20) can be used as a prior probability density.

A.3.2 Quantum Fisher information

The Bures metric (A.20) can also be seen as the quantum equivalent of the Fisher information metric [BZ06]. In terms of the variation of coordinate parameters (A.20) reads,

$$[\mathscr{L}(\rho, \rho + d\rho)]^2 = \frac{1}{2} \mathrm{tr} \left\{ d_{\Theta_\mu} \rho L_\nu \right\} d\Theta_\nu d\Theta_\nu \tag{A.25}$$

A.3 Bures metric

where $\mathrm{d}_{\Theta_\mu}\rho$ is given by,

$$\frac{1}{2}\left(\rho L_\mu + L_\mu \rho\right) = \mathrm{d}_{\Theta_\mu}\rho. \tag{A.26}$$

Hence, we obtain,

$$[\mathscr{L}(\rho, \rho + \mathrm{d}\rho)]^2 = \frac{1}{2}\mathrm{tr}\left\{\frac{1}{2}\left(L_\mu L_\nu + L_\nu L_\mu\right)\rho\right\}\mathrm{d}\Theta_\mu \mathrm{d}\Theta_\nu, \tag{A.27}$$

where the quantum Fisher information metric is identified as,

$$g_{\mu\nu} = \mathrm{tr}\left\{\frac{1}{2}\left(L_\mu L_\nu + L_\nu L_\mu\right)\rho\right\}. \tag{A.28}$$

In its simplest formulation the identification in Eq. (A.27) served as a starting point for the derivation of the quantum speed limit in Eq. (3.58).

A Quantum information theory

B Solution of the parametric harmonic oscillator

In this appendix we summarize the basic concepts and solutions of the parametric harmonic oscillator. We use the method of generating functions, which was introduced by Husimi in 1953 [Hus53]. For a detailed discussion and modern review we refer to [Def08] and [AL10].

B.1 The parametric harmonic oscillator

The Hamiltonian of a quantum mechanical harmonic oscillator with time dependent angular frequency ω_t reads,

$$H_t = \frac{p^2}{2M} + \frac{M}{2}\omega_t^2 x^2 . \tag{B.1}$$

The parameterization ω_t starts at an initial value ω_0 at $t = t_0$ and ends at a final value ω_1 at $t = \tau$. We denote by ϕ_n^t the instantaneous eigenfunctions and by $E_n^t = \hbar\omega_t(n+1/2)$ the instantaneous eigenvalues of the quadratic Hamiltonian (B.1). The dynamics of the harmonic oscillator is Gaussian for any ω_t. By introducing the Gaussian wave function ansatz,

$$\psi_t(x) = \exp\left(\frac{i}{2\hbar}\left[a_t x^2 + 2b_t x + c_t\right]\right), \tag{B.2}$$

B Solution of the parametric harmonic oscillator

the Schrödinger equation for the oscillator can be reduced to a system of three coupled differential equations for the time-dependent coefficients a_t, b_t and c_t,

$$\frac{1}{M} d_t a_t = -\frac{1}{M^2} a_t^2 - \omega_t^2, \qquad (B.3a)$$

$$d_t b_t = -\frac{1}{M} a_t b_t, \qquad (B.3b)$$

$$d_t c_t = \frac{i\hbar}{M} a_t - \frac{1}{M} b_t^2. \qquad (B.3c)$$

The nonlinear equation (B.3a) is of the Riccati type and is, therefore, solvable. It can be mapped to the equation of motion of a classical time-dependent harmonic oscillator via $a_t = M \dot{X}_t / X_t$, and we obtain,

$$d_t^2 X_t + \omega_t^2 X_t = 0. \qquad (B.4)$$

With the solutions of (B.3) the Gaussian wave function $\psi_t(x)$ (B.2) is fully characterized by the time-dependence of the angular frequency ω_t. The general form of the propagator can be determined from $\psi_t(x)$ by noting that

$$\psi_t(x) = \int dx_0 \, U_{t,t_0}(x|x_0) \, \psi_{t_0}(x_0). \qquad (B.5)$$

It is explicitly given by [Hus53],

$$U_{t,t_0} = \sqrt{\frac{M}{2\pi i\hbar X_t}} \exp\left(\frac{iM}{2\hbar X_t}\left[\dot{X}_t x^2 - 2xx_0 + Y_t x_0^2\right]\right), \qquad (B.6)$$

where X_t and Y_t are solutions of Eq. (B.4) satisfying the boundary conditions $X_0 = 0$, $\dot{X}_0 = 1$ and $Y_0 = 1$, $\dot{Y}_0 = 0$, the latter being an expression of the commutation relation between position and momentum.

B.2 Method of generating functions

The time variation of the angular frequency (B.1) induces transitions between different energy eigenstates of the oscillator. We are, thus, interested in the transition probabilities, $p_{m,n}^\tau$, from an initial state, $|n\rangle$ at $t_0 = 0$, to a final state, $|m\rangle$ at $t = \tau$. In

B.2 Method of generating functions

the following, we use the method of generating functions to evaluate $p_{m,n}^\tau$ [Hus53]. We start with the definition,

$$p_{m,n}^\tau = \left| \int dx_0 \int dx\, \phi^{*\tau}_m(x)\, U_{\tau,0}(x|x_0)\, \phi^0_n(x_0) \right|^2, \tag{B.7}$$

and denote the complex conjugate of a number z by z^*. The quadratic generating function of $\phi_n^t(x)$ is given by,

$$\sum_{n=0}^{\infty} u^n\, \phi_n^t(x)\, \phi^{*t}_n(x_0) = \sqrt{\frac{M\omega_t}{\hbar\pi(1-u^2)}} \exp\left(-\frac{M\omega_t}{\hbar}\frac{(1+u^2)(x^2+x_0^2) - 4uxx_0}{2(1-u^2)}\right), \tag{B.8}$$

which can be calculated by a Fourier expansion of the left-hand side of Eq. (B.8). The generating function of $p_{m,n}^\tau$ is, then, defined as,

$$P(u,v) = \sum_{m,n} u^m v^n p_{m,n}^\tau. \tag{B.9}$$

Combining Eqs. (B.7) and (B.8), we find that

$$P(u,v) = \frac{\sqrt{2}}{\sqrt{Q^*(1-u^2)(1-v^2) + (1+u^2)(1+v^2) - 4uv}}. \tag{B.10}$$

The (u,v)-dependence of the generating function, $P(u,v)$, remains the same for all possible transformations ω_t. Details about the specific parameterization of the angular frequency only enter through different numerical values of the parameter Q^* given by,

$$Q^* = \frac{1}{2\omega_0\omega_1}\left\{\omega_0^2\left[\omega_1^2 X_\tau^2 + \dot{X}_\tau^2\right] + \left[\omega_1^2 Y_\tau^2 + \dot{Y}_\tau^2\right]\right\}. \tag{B.11}$$

From a physical point of view, Q^* can be regarded as a measure of the degree of adiabaticity of the process and will be discussed in more detail in the following section. Among the properties of the generating function, $P(u,v)$, it is worth mentioning that the law of total probability, $\sum_n p_{m,n}^\tau = 1$, is fulfilled and is equivalent to,

$$P(u,1) = \frac{1}{1-u}. \tag{B.12}$$

B Solution of the parametric harmonic oscillator

For a constant frequency, $\omega_t \equiv \omega_0$, we note that the solutions of Eq. (B.4) are given by,

$$X_t = \frac{1}{\omega_0}\sin(\omega_0 t), \quad \text{and} \quad Y_t = \cos(\omega_0 t). \tag{B.13}$$

The latter imply with Eq. (B.11) that $Q^* = 1$ and Eq. (B.10) thus simplifies to,

$$P(u,v)\big|_{Q^*=1} = \frac{1}{1-uv}, \tag{B.14}$$

which is equivalent to $p_{m,n}^\tau = \delta_{m,n}$, indicating the absence of transitions, as expected. The symmetry relation of the generating function (B.10), $P(-u,-v) = P(u,v)$, further shows that $p_{m,n}^\tau = 0$ if m, n are of different parity. This is an expression of a selection rule $m = n \pm 2k$, where k is an integer. We mention in addition that the transition probabilities are symmetric, $p_{m,n}^\tau = p_{n,m}^\tau$, following $P(u,v) = P(v,u)$. Explicit expressions for the transition probabilities, $p_{m,n}^\tau$, are given in terms of hypergeometric functions in section B.4.

B.3 Measure of adiabaticity

The parameter Q^* defined in (B.11) can be given a simple physical meaning [Hus53]. We base our discussion of adiabaticity on the equivalent classical harmonic oscillator (B.4) since the generating function, $P(u,v)$ in Eq. (B.10), is fully determined through the classical solutions X_t and Y_t. For an adiabatic transformation, the action of the oscillator, given by the ratio of the energy to the angular frequency, is a time-independent constant. For quasistatic processes we have the two adiabatic invariants,

$$\frac{\dot{X}_t^2 + \omega_t^2 X_t^2}{\omega_t} = \frac{1}{\omega_0}, \quad \text{and} \quad \frac{\dot{Y}_t^2 + \omega_t^2 Y_t^2}{\omega_t} = \omega_0. \tag{B.15}$$

From the definition (B.11) of the parameter Q^*, we see that in this case we simply have $Q^* = 1$. As mentioned earlier, this implies $P(u,v) = (1-uv)^{-1}$ and $p_{m,n}^\tau = \delta_{m,n}$. The latter is an expression of the quantum adiabatic theorem: For infinitely slow transformations no transitions between different quantum states occur. For fast transformations, on the other hand, we can regard Q^* as a measure of the degree of nonadiabaticity of the process. As an illustration, we evaluate mean and

B.3 Measure of adiabaticity

variance of the energy of the oscillator at time τ and express them as a function of Q^*. For a transition from initial state $|n\rangle$ to final state $|m\rangle$, the mean-quantum number of the final state $\langle m \rangle_n$ can be obtained by taking the first derivative of the generating function (B.10) of $p_{m,n}^\tau$,

$$\sum_n u^n \sum_m m\, p_{m,n}^\tau = \partial_v P(u,v)\big|_{v=1} = \frac{Q^*(1+u)-(1-u)}{2(1-u)^2}, \quad \text{(B.16)}$$

and expanding the left hand side of (B.16) in powers of u:

$$\langle m \rangle_n = \sum_m m\, p_{m,n}^\tau = \left(n+\frac{1}{2}\right) Q^* - \frac{1}{2}. \quad \text{(B.17)}$$

Noting that $p_n^0 = \exp(-\beta E_n^0)/Z_0$, the mean energy of the oscillator at time τ then reads

$$\langle H_\tau \rangle = \sum_n \hbar\omega_1 \left(\langle m \rangle_n + \frac{1}{2}\right) p_n^0 = \frac{\hbar\omega_1}{2} Q^* \coth\left(\frac{\beta}{2}\hbar\omega_0\right). \quad \text{(B.18)}$$

Since $\langle m \rangle_n \geq 0$, and, hence, $\langle H_\tau \rangle \geq \hbar\omega_1/2$, the parameter Q^* necessarily satisfies $Q^* \geq 1$ for generic processes. In the zero temperature limit, Eq. (B.18) reduces to,

$$\langle H_\tau \rangle = \frac{\hbar\omega_1}{2} Q^* \quad \text{(B.19)}$$

The above equation corrects a misprint appearing in Eq. (5.21) of Ref. [Hus53] (ω_0 should be replaced by ω_1). The mean-square quantum number, $\langle m^2 \rangle_n - \langle m \rangle_n^2$, at time τ can be calculated in a similar way by considering $\langle m(m-1) \rangle_n$. By differentiating Eq. (B.10) twice, we have,

$$\sum_n u^n \sum_m m(m-1)\, p_{m,n}^\tau = \frac{1+(u-6)u + 3Q^{*2}(u+1)^2 + 4Q^*(u^2-1)}{4(u-1)^3}, \quad \text{(B.20)}$$

and a series expansion in powers of u leads to,

$$\langle m(m-1) \rangle_n = \frac{1}{4}\left[1 - 2n(n+1) - 4Q^*(2n+1) + 3Q^{*2}\left(2n^2+2n+1\right)\right]. \quad \text{(B.21)}$$

B Solution of the parametric harmonic oscillator

The mean-square quantum number is then obtained by combining Eqs. (B.17) and (B.20):

$$\langle m^2 \rangle_n - \langle m \rangle_n^2 = \frac{1}{2}\left(Q^{*2} - 1\right)\left(n^2 + n + 1\right). \tag{B.22}$$

From Eqs. (B.17), (B.18) and (B.22), we can finally write the variance of the energy as,

$$\langle H_\tau^2 \rangle - \langle H_\tau \rangle^2 = \frac{\hbar^2 \omega_1^2}{4} \operatorname{csch}^2\left(\frac{\beta}{2}\hbar\omega_0\right)\left[Q^{*2} + \left(Q^{*2} - 1\right)\cosh(\beta\hbar\omega_0)\right]. \tag{B.23}$$

The zero-temperature limit,

$$\langle H_\tau^2 \rangle - \langle H_\tau \rangle^2 = \frac{\hbar^2 \omega_1^2}{2}\left(Q^{*2} - 1\right), \tag{B.24}$$

is again the correct version of Eq. (5.22) of Ref. [Hus53]. Equation (B.22) indicates that the parameter Q^* directly controls the magnitude of the variance of the occupation number, $\langle m^2 \rangle_n - \langle m \rangle_n^2$. In the adiabatic limit, where $Q^* = 1$, we readily get $\langle m \rangle_n = n$ and $\langle m^2 \rangle_n - \langle m \rangle_n^2 = 0$. We, therefore, recover that for adiabatic processes the system remains in its initial state, $|m\rangle = |n\rangle$. On the other hand, for fast nonadiabatic processes, the mean, $\langle m \rangle_n$, and the dispersion, $\langle m^2 \rangle_n - \langle m \rangle_n^2$, increase with increasing values of Q^*, indicating that the quantum oscillator ends in a final state, $|m\rangle$, which is farther and farther away from the initial state, $|n\rangle$. The latter corresponds to larger and larger final values of the mean energy and energy variance, Eqs. (B.18) and (B.22).

It is worth mentioning that the above discussion of the adiabaticity parameter Q^* for the parametric oscillator is close in spirit to the Einstein criteria for adiabatic processes [Kul57]. Einstein noted that for an adiabatic process, the classical action of the oscillator, $\langle H_t \rangle / \omega_t$, should remain constant and the number of quanta should, therefore, remain unchanged. In the present situation, we have $\langle H_\tau \rangle / \omega_\tau \propto Q^*$, and the action only remains constant when $Q^* = 1$. The latter is precisely the condition that we derived for an adiabatic transformation. For nonadiabatic processes, the parameter $Q^* > 1$, thus, gives a measure for the increase of the classical action of the oscillator. Further discussions of adiabatic measures can be found in Ref. [Tak92].

B.4 Exact transition probabilities

We, here, collect the analytical expressions of the transition probabilities $p_{m,n}^\tau$ [Hus53]. Despite its apparent simplicity, the generation function, $P(u,v)$ in (B.10), cannot be expanded in powers of u and v in an exact series. We, thus, make use of the $p_{m,n}^\tau$ as defined by the matrix elements of the propagator $U_{\tau,0}(x|x_0)$, $p_{m,n}^\tau = |U_{m,n}^\tau|^2$ (B.6). This matrix elements are given by,

$$U_{m,n}^\tau = \int dx_0 \int dx\, \phi^{*\tau}_m(x) U_{\tau,0}(x|x_0) \phi_n^0(x_0). \tag{B.25}$$

We use again the method of generating functions here the linear generating function of $\phi_n^t(x)$ [DL77a],

$$\sum_{n=1}^\infty \left(\frac{\sqrt{\pi} 2^n}{n!}\right)^{1/2} z^n \phi_n^t(x) = \sqrt[4]{\frac{M\omega_t}{\hbar}} \exp\left(-\frac{M\omega_t}{2\hbar} x^2 + 2\sqrt{\frac{M\omega_t}{\hbar}} zx - z^2\right), \tag{B.26}$$

to evaluate the generating function of the propagator,

$$U(u,v) = \sum_{m,n} \left(\frac{\pi 2^{n+m}}{n!\,m!}\right)^{1/2} u^n v^m U_{m,n}^\tau. \tag{B.27}$$

By introducing the complex parameters,

$$\zeta = \omega_1\omega_0 X_\tau - \omega_0 i\dot{X}_\tau + \omega_1 i Y_\tau + \dot{Y}_\tau, \quad |\zeta|^2 = 2\omega_0\omega_1(Q^*-1) \tag{B.28a}$$
$$\sigma = \omega_1\omega_0 X_\tau - \omega_0 i\dot{X}_\tau - \omega_1 i Y_\tau - \dot{Y}_\tau, \quad |\sigma|^2 = 2\omega_0\omega_1(Q^*+1) \tag{B.28b}$$

we can write

$$U(u,v) = \frac{\sqrt[4]{\omega_0\omega_1}}{\sqrt{i\sigma/2\pi}} \exp\left(\frac{\zeta u^2 - 4i\sqrt{\omega_0\omega_1}uv + \zeta^* v^2}{\sigma}\right). \tag{B.29}$$

The matrix elements, $U_{m,n}^\tau$, can then be obtained by a series expansion of (B.28) in powers of u and v [Hus53],

$$U_{m,n}^\tau = \sqrt[4]{2\omega_0\omega_1}\sqrt{\frac{n!\,m!\,\zeta^n \zeta^{*m}}{2^{n+m-1} i\sigma^{n+m+1}}} \sum_{l=0}^{\min(m,n)} \frac{[-2i\sqrt{2/(Q^*-1)}]^l}{l!\,[(n-l)/2]!\,[(m-l)/2]!}. \tag{B.30}$$

B Solution of the parametric harmonic oscillator

According to the selection rule $m = n \pm 2k$, l runs over even numbers only, if m, n are even, and over odd numbers only, if m, n are odd. The explicit expression for the matrix elements, $U^\tau_{m,n}$, then reads for even elements,

$$U^\tau_{2\mu,2\nu} = \sqrt{\frac{2\nu!2\mu!}{2^{2\nu+2\mu-1}i}} \sqrt{\frac{\zeta^{2\nu}\zeta^{*2\mu}}{\sigma^{2\nu+2\mu+1}}} \frac{\sqrt[4]{2\omega_0\omega_1}}{\Gamma(\mu+1)\Gamma(\nu+1)} \\ \times {}_2F_1\left(-\mu, -\nu; \frac{1}{2}; \frac{2}{1-Q^*}\right), \quad (B.31)$$

and for odd elements,

$$U^\tau_{2\mu+1,2\nu+1} = \sqrt{\frac{8(2\nu+1)!(2\mu+1)!}{(1-Q^*)2^{2\nu+2\mu+1}}} \sqrt{\frac{\zeta^{2\nu+1}\zeta^{*2\mu+1}}{\sigma^{2\nu+2\mu+1}}} \frac{\sqrt[4]{2\omega_0\omega_1}}{\Gamma(\mu+1)\Gamma(\nu+1)} \\ \times {}_2F_1\left(-\mu, -\nu; \frac{3}{2}; \frac{2}{1-Q^*}\right). \quad (B.32)$$

We have here introduced the hypergeometric function ${}_2F_1$ [AS72] in order to simplify the sums and write the matrix elements, $U^\tau_{m,n}$, in closed form. Furthermore, $\Gamma(x)$ denotes the Euler Gamma function. Combining everything, we get the explicit expressions for the transition probabilities, which reads for even transitions,

$$p^\tau_{2\mu,2\nu} = \frac{2^{1/2}}{(Q^*+1)^{1/2}} \left(\frac{Q^*-1}{Q^*+1}\right)^{\mu+\nu} \frac{\Gamma(1/2+\mu)\Gamma(1/2+\nu)}{\pi\Gamma(1+\mu)\Gamma(1+\nu)} \\ \times \left[{}_2F_1\left(-\mu, -\nu; \frac{1}{2}; \frac{2}{1-Q^*}\right)\right]^2, \quad (B.33)$$

and for odd transitions,

$$p^\tau_{2\mu+1,2\nu+1} = \frac{2^{7/2}}{(Q^*+1)^{3/2}} \left(\frac{Q^*-1}{Q^*+1}\right)^{\mu+\nu} \frac{\Gamma(3/2+\mu)\Gamma(3/2+\nu)}{\pi\Gamma(1+\mu)\Gamma(1+\nu)} \\ \left[{}_2F_1\left(-\mu, -\nu; \frac{3}{2}; \frac{2}{1-Q^*}\right)\right]^2. \quad (B.34)$$

C Stochastic path integrals

The last appendix is dedicated to an introduction to stochastic Wiener path integrals. Here, the phrase path integrals refers to the generalization of the integral calculus to functionals. A lucidly written introduction to the topic may also be found by Wiegel [Wie86]. The development of a calculus for functionals was initiated by Volterra [Vol65]. His main contribution was a recipe how to handle functions with infinitely many variables. (a) Replace the functional by a function of a finite number, N of variables. (b) Perform all calculations with this function. (c) Take the limit N to infinity. This recipe led after several developments to Feynman's representation of the propagator of the Schrödinger equation by the complex-valued path integral. The Schrödinger equation, on the other hand, takes the form of a general Fokker-Planck equation with complex time. Hence, path integrals are also applicable for the evaluation of the solution of stochastic problems. In the following we review the definition and basic properties of stochastic path integrals, also called Wiener functional integrals.

C.1 Definition and basic properties

Let us start with a very simple problem, namely the free diffusion of a Brownian particle. Its dynamics are described by the diffusion equation,

$$\partial_t p(x,t) = D \partial_x^2 p(x,t), \qquad (C.1)$$

where, as usual, $p(x,t)$ denotes the probability distribution of the particle and D is the diffusion coefficient. Now, let $p_0(x,t)$ be a solution of (C.1) with deterministic initial position, (x_0, t_0),

$$p_0(x, t_0) = \delta(x - x_0). \qquad (C.2)$$

C Stochastic path integrals

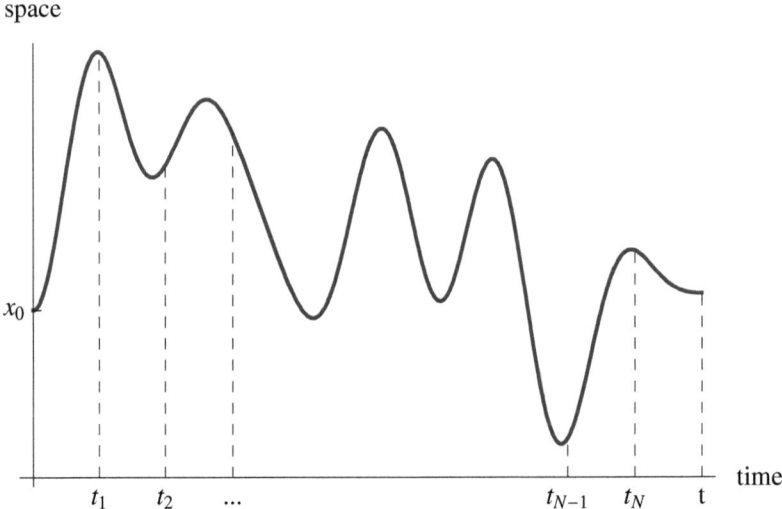

Figure C.1: Particle trajectory which starts at position x_0 at time t_0 and passes trough $N+1$ intermediate time steps, $t_1, t_2, ..., t_N, t$

Hence, the solution $p_0(t,x)$ is evaluated to read,

$$p_0(x,t) = \frac{1}{\sqrt{4\pi D (t-t_0)}} \exp\left(-\frac{(x-x_0)^2}{4D(t-t_0)}\right), \tag{C.3}$$

as can be verified by substitution. The latter solution in Eq. (C.3) is, usually, called the Green's function of the diffusion equation, or the propagator of a Brownian particle. Following Volterra's recipe we have to replace the continuous solution $p_0(t,x)$ by a finite sequence. To this end, we divide the time interval $[t_0, t]$ into $N+1$ equal intervals of length ε. Thus, we obtain an ordered sequence of time points, $t_1 < t_2 < ... < t_N$. Now, we ask for the probability to find a particle, which started in x_0 at t_0, in small neighborhoods around the positions $x_1, x_2, ..., x_N, x$ at times $t_1, t_2, ..., t_N, t$. The neighborhoods are understood as small spatial uncertainties denoted by $dx_1, dx_2, ..., dx_N, dx$. Since we can assume the dynamics in each subinterval to be independent, i.e. Markovian dynamics, the total probability is

C.1 Definition and basic properties

given by the product of the propagator (C.3) over the successive subintervals,

$$(4\pi D\varepsilon)^{-(N+1)/2} \exp\left(-\frac{1}{4D\varepsilon}\sum_{j=0}^{N}(x_{j+1}-x_j)^2\right)\prod_{j=1}^{N+1} dx_j, \tag{C.4}$$

where we defined the final time point $x_{N+1} = x$. In the continuous limit of infinitely many intermediate time steps, $\varepsilon \to 0, N \to \infty, (N+1)\varepsilon = t - t_0$, the latter expression (C.4) converges to the probability for the particle to follow a particular path $X = \{x\}_{t_0}^{t}$ from x_0 to x. Moreover, the exponential in (C.4) can be written in the continuous limit as,

$$\exp\left(-\frac{1}{4D}\int_{t_0}^{t} ds\,(\partial_s x)^2\right). \tag{C.5}$$

By construction one rediscovers after integration of expression (C.4) over all intermediate coordinates, $x_1, x_2, ..., x_N$, the original probability density (C.3). The result is the derivation of the propagator of a Brownian particle as a Wiener path integral,

$$p_0(x,t) = \int_{x_0}^{x} \mathcal{D}X \exp\left(-\frac{1}{4D}\int_{t_0}^{t} ds\,(\partial_s x)^2\right). \tag{C.6}$$

It is worth emphasizing that the right-hand side of the latter equation (C.6) is merely a notation for Volterra's recipe applied to a particular functional. The symbol $\mathcal{D}X$ denotes the *infinitesimal small* collection of sequences $X = \{x\}_{t_0}^{t}$, which obey,

$$\begin{aligned} x(t_0) &= x_0, \\ x_1 &< x(t_1) < x_1 + dx_1, \\ x_2 &< x(t_2) < x_2 + dx_2, \\ &\dots, \\ x_N &< x(t_N) < x_N + dx_N, \\ x(t) &= x. \end{aligned} \tag{C.7}$$

C Stochastic path integrals

Hence, the symbol $\mathscr{D}X$ can equivalently be expressed as a product over all intermediate coordinates,

$$\int_{x_0}^{x} \mathscr{D}X = \lim_{N\to\infty} (4\pi D\varepsilon)^{-(N+1)/2} \prod_{j=1}^{N+1} \int_{-\infty}^{+\infty} \mathrm{d}x_j . \tag{C.8}$$

Further, on the left-hand side of Eq. (C.8) the boundaries of the path integral indicate, that all possible paths, X, start and end in the same initial and final coordinates, x_0 and x, respectively. On the right hand side, however, the integrals runs over the whole probability space. For the evaluation of path integrals one always has to perform all integrations in the discrete formulation and, then, take the limit $N \to \infty$.

C.2 Onsager-Machlup functional for space dependent diffusion

In the latter section we considered the simplest, physically relevant case, namely the diffusion equation of a free Brownian particle. Next, we turn to the more complicated problem of general Fokker-Planck equations. A detailed derivation of the according path integrals was proposed by Grabert and Green [GG79], whereas we follow, here, the more heuristic approach of Risken [Ris89]. For the sake of generality, we consider a generic driven Fokker-Planck equation, with time and position dependent drift and diffusion coefficients, D_1 and D_2, of the form,

$$\partial_t p(x,t) = \mathscr{F}(x,t) p(x,t) , \tag{C.9}$$

where the linear operator $\mathscr{F}(x,t)$ is given by,

$$\mathscr{F}(x,t) = -\partial_x D_1(x,t) + \partial_x^2 D_2(x,t) . \tag{C.10}$$

A formal solution of Eq. (C.9) with the deterministic initial distribution (C.2) can be written as,

$$p_0(x,t) = \mathscr{T}_{>} \exp\left(\mathscr{F}(x,t)(t-t_0)\right) \delta(x-x_0) , \tag{C.11}$$

where $\mathscr{T}_{>}$ is the time ordering operator. The time ordering has to be included in the solution in order to formulate the corresponding fundamental system of Eq. (C.9)

C.2 Onsager-Machlup functional for space dependent diffusion

as a well-defined formulation. For small increments of time, $\varepsilon = t - t_0$, the latter solution $p_0(x,t)$ can be expanded to read,

$$p_\varepsilon(x,t) \simeq \left[1 + \mathscr{F}(x,t)\varepsilon + \mathcal{O}\left(\varepsilon^2\right)\right]\delta(x-x_0). \tag{C.12}$$

From the latter expansion we conclude that for small times ε the time ordering can be omitted. Thus, we obtain up to corrections of the order ε^2,

$$p_\varepsilon(x,t) = \left(1 - \partial_x D_1(x_0,t)\varepsilon + \partial_x^2 D_2(x_0,t)\varepsilon\right)\delta(x-x_0), \tag{C.13}$$

where we replaced x by x_0 in the drift and diffusion coefficients. If we now introduce the representation of the δ-function in terms of a Fourier integral, we obtain for small ε,

$$p_\varepsilon(x,t) = \left(1 - \partial_x D_1(x_0,t)\varepsilon + \partial_x^2 D_2(x_0,t)\varepsilon\right)\frac{1}{2\pi}\int du \exp(iu(x-x_0)). \tag{C.14}$$

By further evaluating the derivatives in Eq. (C.14) we then obtain,

$$\begin{aligned}p_\varepsilon(x,t) &= \frac{1}{2\pi}\int du \exp(iu(x-x_0))\left(1 - iuD_1(x_0,t)\varepsilon - u^2 D_2(x_0,t)\varepsilon\right) \\ &\simeq \frac{1}{2\pi}\int du \exp\left(iu(x-x_0) - iuD_1(x_0,t)\varepsilon - u^2 D_2(x_0,t)\varepsilon\right).\end{aligned} \tag{C.15}$$

Thus, the propagator of a general Fokker-Planck equation is given for small time steps ε as,

$$p_\varepsilon(x,t) = \frac{1}{\sqrt{4\pi D_2(x_0,t)\varepsilon}}\exp\left(-\frac{(x-x_0-D_1(x_0,t)\varepsilon)^2}{4D_2(x_0,t)\varepsilon}\right). \tag{C.16}$$

As before (cf. Eq. (C.4)) the probability for a sequence of N successive, independent steps is given by the product of the propagator of the individual, infinitesimal steps,

$$\begin{aligned}p_0(x,t) = \lim_{N\to\infty} &\prod_{j=1}^{N+1}\left(4\pi D_2(x_j,t)\varepsilon\right)^{-(N+1)/2} \\ &\times \int_{-\infty}^{+\infty} dx_j \exp\left(-\sum_{j=1}^{N}\frac{(x_{j+1}-x_j-D_1(x_j,t_j)\varepsilon)^2}{4D_2(x_j,t_j)\varepsilon}\right).\end{aligned} \tag{C.17}$$

C Stochastic path integrals

The latter equation can, then, be written by taking the continuous limit as,

$$p_0(x,t) = \int_{x_0}^{x} \mathscr{D}X \exp\left(-\int_{t_0}^{t} ds\, S(x_s, \dot{x}_s)\right), \tag{C.18}$$

where we introduced the generalized Onsager-Machlup functional, i.e. the stochastic action,

$$S(x_t, \dot{x}_t) = \frac{[\dot{x}_t - D_1(x_t, t)]^2}{4 D_2(x_t, t)}. \tag{C.19}$$

The form of $S(x_t, \dot{x}_t)$, however, is not unique. A class of statistically equivalent forms can be derived by rearranging the coordinates and derivatives in Eqs. (C.13) and (C.14). A more rigorous analysis is presented in [CJ06]. For our purpose we reformulate Eq. (C.14) by keeping the replacement of x and x_0 in the drift coefficient, whereas the diffusion coefficient remains unchanged. Thus, we obtain,

$$\begin{aligned}\tilde{p}_\varepsilon(x,t) &= \left(1 - \partial_x D_1(x_0,t)\,\varepsilon + \partial_x^2 D_2(x,t)\,\varepsilon\right)\delta(x-x_0)\\ &= \delta(x-x_0) - D_1(x_0,t)\,\varepsilon\,\delta'(x-x_0)\\ &\quad + \partial_x \left(D_2'(x,t)\,\delta(x-x_0) + D_2(x,t)\,\delta'(x-x_0)\right)\varepsilon,\end{aligned} \tag{C.20}$$

where the prime is a short notation for the partial derivative with respect to space. After evaluating the first derivative in the diffusion term we replace x by x_0 and perform the second derivative afterwards. Hence, we obtain the statically equivalent formulation of Eq. (C.15) as,

$$\begin{aligned}\tilde{p}_\varepsilon(x,t) &= \frac{1}{2\pi}\int du \exp(iu(x-x_0))\\ &\quad \times \left[1 - iu\left(D_1(x_0,t) - D_2'(x_0,t)\right)\varepsilon - u^2 D_2(x_0,t)\,\varepsilon\right]\\ &\simeq \frac{1}{2\pi}\int du \exp\left[iu(x-x_0) - iu\left(D_1(x_0,t) - D_2'(x_0,t)\right)\varepsilon - u^2 D_2(x_0,t)\,\varepsilon\right].\end{aligned} \tag{C.21}$$

Finally, by evaluation of the Fourier integral, a statistically equivalent propagator of a general Fokker-Planck equation for small time steps, ε, is given by,

$$\tilde{p}_\varepsilon(x,t) = \frac{1}{\sqrt{4\pi D_2(x_0,t)\varepsilon}} \exp\left(-\frac{[x - x_0 - (D_1(x_0,t) - D_2'(x_0,t))\,\varepsilon]^2}{4 D_2(x_0,t)\varepsilon}\right). \tag{C.22}$$

C.2 Onsager-Machlup functional for space dependent diffusion

Accordingly, the continuous limit results in the modified, but equivalent stochastic action functional,

$$\widetilde{S}(x_t, \dot{x}_t) = \frac{\left[\dot{x}_t - \left(D_1(x_t,t) - D'_2(x_t,t)\right)\right]^2}{4 D_2(x_t,t)}. \tag{C.23}$$

It is remarkable, that the latter Onsager-Machlup functional (C.23) merely has a modified drift, whereas the path discretization and, hence, the increment $\mathscr{D}X$ remain unchanged. The form in Eq. (C.23) of the stochastic action is appropriate for our purposes and, thus, used in the present thesis. The here proposed derivation is heuristic in so far as we used some degree of freedom how to interpret the δ-functions in Eqs. (C.20) and (C.21). A rigorous treatment can be found in the paper of Grabert and Green [GG79].

C Stochastic path integrals

D Acknowledgments

Zu guter Letzt möchte ich die Gelegenheit wahrnehmen allen zu danken, die an der Erstellung der vorliegenden Dissertation einen Anteil hatten. Allen voran bedanke ich mich bei meinem Doktorvater Dr. Eric Lutz. Durch seine langjährige und intensive Förderung leistete er den wesentlichen Beitrag zu meiner wissenschaftlichen Entwicklung. Des Weiteren bedanke ich mich bei Prof. Dr. Peter Hänggi, der auf Grund seiner klaren und direkten Fragestellungen ein wissenschaftliches Vorbild ist, für seine Förderung und für das Erstellen eines Gutachtens über die vorliegende Arbeit. Für ein weiteres Gutachten bedanke ich mich bei Prof. Dr. Udo Seifert. Es war auch immer eine Freude, auf internationalen Konferenzen mit ihm zu diskutieren. Ich bedanke mich bei Prof. Dr. Ulrich Eckern dafür, dass er mich zur Zeit meiner Vordiplomsprüfungen überzeugen konnte, der Physik treu zu bleiben, und für seine Teilnahme an meiner Prüfungskommission. Schließlich danke ich Prof. Ferdinand Schmidt-Kaler für die Übernahme des experimentellen Teils meiner Prüfung und für unsere erfolgreiche wissenschaftliche Zusammenarbeit. In diesem Zusammenhang bedanke ich mich auch insbesondere bei Dr. Gerhard Huber, mit dem ich viele lehrreiche Gespräche führen konnte. Nicht unerwähnt lassen möchte ich auch Prof. Peter Talkner und Prof. Dr. Gert-Ludwig Ingold für ihre offenen Türen und das Teilen ihres Wissens und ihrer Erfahrung mit mir. Auch bedanke ich mich bei Prof. Dr. Dirk Blömker, der den größten Teil meiner mathematischen Ausbildung leistete. Für eine weitere Zusammenarbeit zwischen Theorie und Experiment bedanke ich mich bei Dr. Roland Schäfer und Christoph Kaiser. Ein besonderer Dank gilt meinen Studenten Michael Brunner, Malte Lehmann und Obinna Abah, deren Betreuung ich während der letzten Jahre zum Teil übernehmen durfte, und die alle mit einem Teil ihrer Ergebnisse in dieser Arbeit vertreten sind. Einen besonderen Stellenwert in meiner Ausbildung nehmen der Besuch von zwei Sommerschulen ein. Ich bedanke mich bei all meinen Kommilitonen, die ich dort traf und mit denen ich bis lange in die Nacht diskutierte. Besonders hervorheben möchte ich Dr. Klaus Röller, der mit mir in Paris war, und Jeroen Devreese, mit dem ich zwei Wochen in einem Kloster

D Acknowledgments

nahe Leuven verbrachte. Ein ganz besonderer Dank gilt meinem beiden Kollegen und Freunden Georg Reuther und Peter Siegle. In den letzten Monaten, in denen wir zeitgleich unsere Dissertationen fertig stellten, verbrachten wir viel Zeit damit uns gegenseitig zu motivieren und zu inspirieren. Nicht vergessen möchte ich auch Mathias Keiß, der mich bei der Gestaltung der gedruckten Version unterstützte. Ganz besonders bedanke ich mich auch bei Birgit Kießig für die Zeit, die sie aufwendete um meine Dissertation auf Fehler zu überprüfen. Mit der wichtigste Dank geht an meine Eltern, Isabella und Alfred Deffner, und meinen Bruder, Christoph Deffner, die immer an mich glaubten, mich immer wieder aufrichteten und in allen Entscheidungen hinter mir standen.

Bibliography

[AA85] J. Nulton, P. Salamon, B. Andresen, and Qi Anmin, *Quasistatic processes as step equilibrations*, J. Chem. Phys. **83** (1985), 334.

[AE05] K. M. R. Audenaert and J. Eisert, *Continuity bounds on the quantum relative entropy*, J. Math. Phys. **46** (2005), 102104.

[AG08] J. Ankerhold and H. Grabert, *Erratum: Strong friction limit in quantum mechanics: The quantum Smoluchowski equation*, Phys. Rev. Lett. **101** (2008), 119903.

[AH69] V. Ambegaokar and B. I. Halperin, *Voltage due to thermal noise in the dc Josephson effect*, Phys. Rev. Lett. **22** (1969), 1364.

[AL10] S. Deffner, O. Abah, and E. Lutz, *Quantum work statistics of linear and nonlinear parametric oscillators*, Chem. Phys. **375** (2010), 200.

[Ali02] R. Alicki, *Invitation to quantum dynamical semigroups*, Lecture Notes in Physics **597** (2002), 239.

[Ank01] J. Ankerhold, *Quantum decay rates for driven barrier potentials in the strong friction limit*, Phys. Rev. E **64** (2001), 060102 (R).

[Ank04] _____, *Overdamped quantum phase diffusion and charging effects in Josephson junctions*, Europhys. Lett. **67** (2004), 280.

[AS72] M. Abaramowitz and I. E. Stegun, *Handbook of mathematical functions*, United States department of commerce, Washington D.C., USA, 1972.

[Aul04] B. Aulbach, *Gewöhnliche Differentialgleichungen*, Spektrum, Akademischer Verlag, Elsevier, München, Germany, 2004.

Bibliography

[BC94] S. L. Braunstein and C. M Caves, *Statistical distance and the geometry of quantum states*, Phys. Rev. Lett. **72** (1994), 3439.

[Bek74] J. D. Bekenstein, *Generalized second law of thermodynamics in black-hole physics*, Phys. Rev. D **9** (1974), 3292.

[Bek81] _____, *Universal upper bound on the entropy-to-energy ratio for bounded systems*, Phys. Rev. D **23** (1981), 287.

[BL11] S. Deffner, M. Brunner, and E. Lutz, *Quantum fluctuation theorems in the strong damping limit*, Europhys. Lett. **94** (2011), 30001.

[BM90a] A. P. Prudnikov, Y. A. Brychow, and O. I. Marichev, *Integrals and series*, vol. 1, Gordon and Breach, Amsterdam, The Netherlands, 1990.

[BM90b] _____, *Integrals and series*, vol. 2, Gordon and Breach, Amsterdam, The Netherlands, 1990.

[BP07] H.-P. Breuer and F. Petruccione, *The theory of open quantum systems*, Oxford University Press, New York, USA, 2007.

[Bre67] H. J. Bremermann, *Quantum noise and information*, Proc. Fifth Berkeley Symp. on Math. Statist. and Prob. **4** (1967), 15.

[Bre03] H.-P. Breuer, *Quantum jumps and entropy production*, Phys. Rev. A **68** (2003), 032105.

[Bru10] M. Brunner, *Quantum work distributions for overdamped system far from equilibrium*, Diploma thesis, 2010, University of Augsburg.

[BS90] J. D. Bekenstein and M. Schiffer, *Quantum limitations on the storage and transmission of information*, Int. J. Mod. Phys. C **1** (1990), 355.

[Bur68] D. J. C. Bures, *Tensor products of W^*-algebras*, Pac. J. Math. **27** (1968), 13.

[Bur69] _____, *An extension of Kakutani's theorem on infinite product measures to the tensor product of semifinite W^*-algebras*, Tran. Amer. Math. Soc. **135** (1969), 199.

Bibliography

[BZ06] I. Bengtsson and K. Życzkowski, *Geometry of quantum states*, Cambridge University Press, New York, USA, 2006.

[Car24] S. Carnot, *Réflexions sur la puissance motrice de feu et sur les machines propres à développer cette puissance*, Bachelier, Paris, France, 1824.

[Cer09] G. Cerefolini, *Nanoscale devices*, Springer, Berlin, Germany, 2009.

[CH09] P. Talkner, M. Campisi, and P. Hänggi, *Fluctuation theorems in driven open quantum systems*, J. Stat. Mech. **2009** (2009), P02025.

[CJ06] V. Y. Chernyak, M. Chertkov, and C. Jarzynski, *Path-integral analysis of fluctuation theorems for general Langevin processes*, J. Stat. Mech. **2006** (2006), P08001.

[CL81] A. O. Caldeira and A. J. Leggett, *Influence of dissipation on quantum tunneling in macroscopic systems*, Phys. Rev. Lett. **46** (1981), 211.

[CL83] _____, *Path integral approach to quantum Brownian motion*, Physica A **121** (1983), 587.

[Cla64] R. Clausius, *Abhandlungen über die mechanische Wärmetheorie*, Vieweg, Braunschweig, Germany, 1864.

[CM93] D. Evans, E. G. D. Cohen, and G. Morris, *Probability of second law violations in sharing steady states*, Phys. Rev. Lett. **71** (1993), 2401.

[Cro98] G. E. Crooks, *Nonequilibrium measurements of free energy differences for microscopically reversible Markovian systems*, J. Stat. Phys. **90** (1998), 1481.

[Cro99] _____, *Entropy production fluctuation theorems and the nonequilibrium work relation for free energy differences*, Phys. Rev. E **60** (1999), 2721.

[Cro07] _____, *Measuring thermodynamic length*, Phys. Rev. Lett. **99** (2007), 100602.

Bibliography

[CZ96]   J. F. Poyatos, J. I. Cirac, and P. Zoller, *Quantum reservoir engineering with laser cooled trapped ions*, Phys. Rev. Lett. **77** (1996), 4728.

[CZ03]   A. Fennimore, T. Yuzvinsky, W. Han, M. Fuhrer, J. Cumings, and A. Zettl, *Rotational actuators based on carbon nanotubes*, Nature (London) **424** (2003), 408.

[Dav54]   R. W. Davies, *The connection between the Smoluchowski equation and the Kramers-Chandasekhar equation*, Phys. Rev. **93** (1954), 1169.

[Def08]   S. Deffner, *Quantum work relations*, Diploma thesis, 2008, University of Augsburg.

[dGM84]   S. R. de Groot and P. Mazur, *Non-equilibrium thermodynamics*, Dover Publications, Dover, UK, 1984.

[Dit99]   J. Dittmann, *Explicit formulae for the Bures metric*, J. Phys. A **32** (1999), 2663.

[DK97]   L. F. Cugliandolo, D. S. Dean, and J. Kurchan, *Fluctuation-dissipation theorems and entropy production in relaxational systems*, Phys. Rev. Lett. **79** (1997), 2168.

[DL77a]   C. Cohen-Tannoudji, B. Diu, and F. Laloë, *Quantum mechanics*, vol. 1, Herrmann, Paris, France, 1977.

[DL77b]   ———, *Quantum mechanics*, vol. 2, Herrmann, Paris, France, 1977.

[DL08a]   G. Huber, F. Schmidt-Kaler, S. Deffner, and E. Lutz, *Employing trapped cold ions to verify the quantum Jarzynski equality*, Phys. Rev. Lett. **101** (2008), 070403.

[DL08b]   S. Deffner and E. Lutz, *Nonequilibrium work distribution of a quantum harmonic oscillator*, Phys. Rev. E **77** (2008), 021128.

[DL09]   R. Dillenschneider and E. Lutz, *Quantum Smoluchowski equation for driven systems*, Phys. Rev. E **80** (2009), 042101.

[DL10]   S. Deffner and E. Lutz, *Generalized Clausius inequality for nonequilibrium quantum processes*, Phys. Rev. Lett. **105** (2010), 170402.

Bibliography

[DN99] D. Daems and G. Nicolis, *Entropy production and phase space volume contraction*, Phys. Rev. E **59** (1999), 4000.

[Ein05] A. Einstein, *Über von der molekularkinetischen Theorie der Wärme geforderten Bewegung von in ruhenden Flüßigkeiten suspendierten Teilchen*, Ann. Phys. **17** (1905), 549.

[EM06] M. Esposito and S. Mukamel, *Fluctuation theorems for quantum master equations*, Phys. Rev. E **73** (2006), 046129.

[FK87] G. W. Ford and M. Kac, *On the quantum Langevin equation*, J. Stat. Phys. **46** (1987), 803.

[FO01] G. W. Ford and R. F. O'Connell, *Exact solution of the Hu-Paz-Zhang master equation*, Phys. Rev. D **64** (2001), 105020.

[FV63] R. P. Feynman and F. L. Vernon, *The theory of a general quantum system interacting with a linear dissipative system*, Ann. Phys. (New York) **24** (1963), 118.

[GC95] G. Gallavotti and E. G. D. Cohen, *Dynamical ensembles in nonequilibrium statistical mechanics*, Phys. Rev. Lett. **74** (1995), 2694.

[GC10] M. Murphy, S. Montangero, V. Giovannetti, and T. Calarco, *Communication at the quantum speed limit along a spin chain*, Phys. Rev. A **82** (2010), 022318.

[GG79] H. Grabert and M. S. Green, *Fluctuations and nonlinear irreversible processes*, Phys. Rev. A **19** (1979), 1747.

[GM05] R. Gross and A. Marx, *Applied superconductivity: Josephson effect and superconducting electronics*, lecture notes, 2005.

[Gri91] R. Grigorieff, *A note on von NeumannâĂŹs trace inequality*, Math. Nachr. **151** (1991), 327.

[GS09] T. Caneva, M. Murphy, T. Calarco, R. Fazio, S. Montangero, V. Giovannetti, and G. E. Santoro, *Optimal control at the quantum speed limit*, Phys. Rev. Lett. **103** (2009), 240501.

Bibliography

[HM09] P. Hänggi and F. Marchesoni, *Artificial Brownian motors: controlling transport on the nanoscale*, Rev. Mod. Phys. **81** (2009), 387.

[HS01] T. Hatano and S. Sasa, *Steady-state thermodynamics of Langevin systems*, Phys. Rev. Lett. **86** (2001), 3463.

[Hue92] M. Huebner, *Explicit computation of the Bures distance for density matrices*, Phys. Lett. A **163** (1992), 239.

[Hus53] K. Husimi, *Miscellanea in elementary quantum mechanics, II*, Prog. Theo. Phys. **9** (1953), 381.

[IG99] G. Ingold and H. Grabert, *Effect of zero point phase fluctuations on Josephson tunneling*, Phys. Rev. Lett. **83** (1999), 3721.

[IP98] H. Grabert, G.-L. Ingold, and B. Paul, *Phase diffusion and charging effects in Josephson junctions*, Europhys. Lett. **44** (1998), 360.

[JAP05] H. Grabert, J. Ankerhold, and P. Pechukas, *Quantum Brownian motion with large friction*, Chaos **15** (2005), 026106.

[Jar97] C. Jarzynski, *Nonequilibrium equality for free energy differences*, Phys. Rev. Lett. **78** (1997), 2690.

[Jar08] ———, *Nonequilibrium work relations: foundations and applications*, Eur. Phys. J. B **64** (2008), 331.

[JB02] J. Liphardt, S. Dumont, S. B. Smith, I. Tinoco, Jr., and C. Bustamante, *Equilibrium information from nonequilibrium measurements in an experimental test of Jarzynski's equality*, Science **296** (2002), 1832.

[JK10] P. J. Jones and P. Kok, *Geometric derivation of the quantum speed limit*, Phys. Rev. A **82** (2010), 022107.

[Jos62] B. D. Josephson, *Possible new effects in superconductive tunnelling*, Phys. Lett. **1** (1962), 251.

[Joz94] R. Jozsa, *Fidelity for mixed quantum states*, J. Mod. Opt. **41** (1994), 2315.

Bibliography

[JW05]   C. Jarzynski and D. K. Wójcik, *Classical and quantum fluctuation theorems for heat exchange*, Phys. Rev. Lett. **92** (2005), 230602.

[Kak48] S. Kakutani, *On equivalence of infinite product measures*, Ann. Math. **49** (1948), 214.

[KI04]   A. Golobov, M. Kupriyanov, and E. Il'ichev, *The current-phase relation in Josephson junctions*, Rev. Mod. Phys. **76** (2004), 411.

[KL51]   S. Kullback and R. A. Leibler, *On information and sufficiency*, Ann. Math. Stat. **22** (1951), 79.

[KM65]   G. W. Ford, K. Kac, and P. Mazur, *Statistical mechanics of assemblies of coupled oscillators*, J. Math. Phys. **6** (1965), 504.

[KP07]   R. Avakyan, A. Hayrapetyan, B. Khachatryan, and R. Petrosyan, *Coherent presentation of the density operator of the harmonic oscillator in thermostat*, Phys. Lett. A **372** (2007), 77.

[Kra40] H. Kramers, *Brownian motion in a field of force and the diffusion model of chemical reactions*, Physica (Utrecht) **VII** (1940), 284.

[KS10]   C. Kaiser and R. Schäfer, *(private communication)*, 2010.

[Kub57] R. Kubo, *Statistical-mechanical theory of irreversible processes. I. General theory and simple applications to magnetic and conduction problems*, J. Phys. Soc. Jpn. **12** (1957), 570.

[Kul57] R. M. Kulsrud, *Adiabatic invariant of the harmonic oscillator*, Phys. Rev. **106** (1957), 205.

[Kul78] S. Kullback, *Information theory and statistics*, Peter Smith, Gloucester, USA, 1978.

[Kur98] J. Kurchan, *Fluctuation theorem for stochastic dynamics*, J. Phys. A **31** (1998), 3719.

[Kur00] ———, *A quantum fluctuation theorem*, arXiv:cond-mat/0007360v2., 2000.

Bibliography

[Lan08] P. Langevin, *Sur la théorie du mouvement Brownien*, C. R. Acad. Sci. (Paris) **146** (1908), 530.

[LD10] M. Lehmann and S. Deffner, *NIM Summer Research Program 2010*, 2010.

[LG97] D. S. Lemmons and A. Gythiel, *Paul Langevin's 1908 paper "On the theory of Brownian motion"*, Am. J. Phys. **65** (1997), 1079.

[LH07] P. Talkner, E. Lutz, and P. Hänggi, *Fluctuation theorems: Work is not an observable*, Phys. Rev. E **75** (2007), 050102 (R).

[Lik86] K. K. Likharev, *Dynamics of Josephson junctions and circuits*, Gordon and Breach science publishers, Philadelphia, USA, 1986.

[Lin74] G. Lindblad, *Measurements and information for thermodynamic quantities*, J. Stat. Mech. **11** (1974), 231.

[Lin75] _____, *Completely positive maps and entropy inequalities*, Commun. Math. Phys. **40** (1975), 147.

[Lin76] _____, *On the generators of quantum dynamical semigroups*, Commun. Math. Phys. **48** (1976), 119.

[Lin83] _____, *Non-equilibrium entropy and irreversibility*, D. Reidel Publishing Company, Dordrecht, Holland, 1983.

[LM03] V. Giovannetti, S. Lloyd, and L. Maccone, *Quantum limits to dynamical evolution*, Phys. Rev. A **67** (2003), 052109.

[LP05] A. Majtey, P. Lamberti, and D. Prato, *Jensen-Shannon divergence as a measure of distinguishability between mixed quantum states*, Phys. Rev. A **72** (2005), 052310.

[LS99] J. Lebowitz and H. Spohn, *A Gallavotti-Cohen-type symmetry in the large deviation functional for stochastic dynamics*, J. Stat. Phys. **95** (1999), 333.

Bibliography

[LT09] L. B. Levitin and Y. Toffoli, *Fundamental limit on the rate of quantum dynamics: The unified bound is tight*, Phys. Rev. Lett. **103** (2009), 160502.

[MA10a] S. A. Maier and J. Ankerhold, *Low-temperature quantum fluctuations in overdamped ratchets*, Phys. Rev. E **82** (2010), 021104.

[MA10b] _____, *Quantum Smoluchowski equation: A systematic study*, Phys. Rev. E **81** (2010), 021107.

[Mir75] L. Mirsky, *A trace inequality of John von Neumann*, Monatshefte für Math. **79** (1975), 303.

[ML98] N. Margolus and L. B. Levitin, *The maximum speed of dynamical evolution*, Physica D **120** (1998), 188.

[MT45] L. Mandelstam and I. Tamm, *The uncertainty relation between energy and time in nonrelativistic quantum mechanics*, J. Phys. (USSR) **9** (1945), 249.

[MW00a] C. J. Myatt, B. E. King, Q. A. Turchette, C. A. Sackett, D. Kielpinski, W. M. Itano, C. Monroe, and D. J. Wineland, *Decoherence of quantum superpositions through coupling to engineered reservoirs*, Nature (London) **403** (2000), 269.

[MW00b] Q. Turchette, D. Kielpinski, B. King, D. Leibfried, D. Meekhof, C. Myatt, M. Rowe, C. Sackett, C. Wood, W. Itano, C. Monroe, and D. Wineland, *Heating of trapped ions from the quantum ground state*, Phys. Rev. A **61** (2000), 063418.

[NB84] P. Salamon, J. Nulton, and R. Berry, *Length in statistical thermodynamics*, J. Chem. Phys. **82** (1984), 2433.

[NC00] M . A. Nielsen and I. L. Chuang, *Quantum computation and quantum information*, Cambridge University Press, Cambridge, UK, 2000.

[Ons31a] L. Onsager, *Reciprocal relations in irreversible processes. I.*, Phys. Rev. **37** (1931), 405.

Bibliography

[Ons31b] _____, *Reciprocal relations in irreversible processes. II.*, Phys. Rev. **38** (1931), 2265.

[OT97] T. A. Savard, K. M. O'Hara, and J. E. Thomas, *Laser-noise-induced heating in far-off resonance optical traps*, Phys. Rev. A **56** (1997), R1095.

[OZ01] W. Hebisch, R. Olkiewicz, and B. Zegarlinski, *On upper bound for the quantum entropy*, Lin. Alg. Appl. **329** (2001), 89.

[PdB07] R. Kawai, J. M. R. Parrondo, and C. Van den Broeck, *Dissipation: The phase-space perspective*, Phys. Rev. Lett. **98** (2007), 080602.

[PG01] J. Ankerhold, P. Pechukas, and H. Grabert, *Strong friction limit in quantum mechanics: The quantum Smoluchowski equation*, Phys. Rev. Lett. **87** (2001), 086802.

[Pri47] I. Prigogine, *Etude thermodynamique des phénomènes irréversibles*, Dunod, Paris and Desoer, Liège, Belgium, 1947.

[PZ92] B. L. Hu, J. P. Paz, and Y. Zhang, *Quantum Brownian motion in a general environment: Exact master equation with nonlocal dissipation and colored noise*, Phys. Rev. D **45** (1992), 2843.

[Rao45] C. R. Rao, *Information and the accuracy attainable in the estimation of statistical parameters*, Bulletin of the Calcutta Mathematical Society **37** (1945), 81.

[Rei02] P. Reimann, *Brownian motors: noisy transport far from equilibrium*, Phys. Rep. **361** (2002), 57.

[Ris89] H. Risken, *The Fokker-Planck equation*, Springer, Berlin, Germany, 1989.

[Rup95] G. Ruppeiner, *Riemannian geometry in thermodynamic fluctuation theory*, Rev. Mod. Phys. **67** (1995), 605.

[SB83] P. Salamon and R. Berry, *Thermodynamic length and dissipated availability*, Phys. Rev. Lett. **51** (1983), 1127.

Bibliography

[SB06] V. Blickle, T. Speck, L. Helden, U. Seifert, and C. Bechinger, *Thermodynamics of a colloidal particle in a time-dependent nonharmonic potential*, Phys. Rev. Lett. **96** (2006), 070603.

[Sch66] F. Schlögl, *Zur statistischen Theorie der Entropieproduktion in nicht abgeschlossenen Systemen*, Z. Phys. **191** (1966), 81.

[Sch89] _____, *Probability and heat*, Vieweg, Braunschweig, Germany, 1989.

[Scu98] H. Scutaru, *Fidelity for displaced squeezed thermal states and the oscillator semigroup*, J. Phys. A **31** (1998), 3659.

[SE02] G. Wang, E. Sevick, E. Mittag, D. Searles, and D. Evans, *Experimental demonstration of violations of the second law of thermodynamics for small systems and short time scales*, Phys. Rev. Lett. **89** (2002), 050601.

[SE04] D. M. Carberry, J. C. Reid, G. M. Wang, E. M. Sevick, D. J. Searles, and D. J. Evans, *Fluctuations and irreversibility: An experimental demonstration of a second-law-like theorem using a colloidal particle held in an optical trap*, Phys. Rev. Lett. **92** (2004), 140601.

[Sei05] U. Seifert, *Entropy production along a stochastic trajectory and an integral fluctuation theorem*, Phys. Rev. Lett. **95** (2005), 040602.

[Sei08] _____, *Stochastic thermodynamics: principles and perspectives*, Eur. Phys. J. B **64** (2008), 423.

[Sek98] K. Sekimoto, *Langevin equation and thermodynamics*, Prog. Theo. Phys. Suppl. **130** (1998), 17.

[SI87] H. Grabert, P. Schramm, and G.-L. Ingold, *Localization and anomalous diffusion of a damped quantum particle*, Phys. Rev. Lett. **58** (1987), 1285.

[SI88] _____, *Quantum Brownian motion: The functional integral approach*, Phys. Rep. **168** (1988), 115.

Bibliography

[Spo78] H. Spohn, *Entropy production for quantum dynamical semigroups*, J. Math. Phys. **19** (1978), 1227.

[SS00] S. S. Dragomir, M. L. Scholz, and J. Sunde, *Some upper bounds for relative entropy and applications*, Comp. Math. Appl. **39** (2000), 91.

[SSK08] G. Huber, T. Deuschle, W. Schnitzler, R. Reichle, K. Singer, and F. Schmidt-Kaler, *Transport of ions in a segmented linear Paul trap in printed-circuit-board technology*, New J. Phys. **10** (2008), 013004.

[SZ03] H.-J. Sommers and K. Zyczkowski, *Bures volume of the set of mixed quantum states*, J. Phys. A **36** (2003), 10083.

[Tak92] K. Takayama, *Exact study of adiabaticity*, Phys. Rev. A **45** (1992), 2618.

[Tas00] H. Tasaki, *Jarzynski relations for quantum systems and some applications*, arXiv:cond-mat/0000244v2, 2000.

[TB05] D. Collin, F. Ritort, C. Jarzynski, S. B. Smith, I. Tinoco, and C. Bustamante, *Verification of the Crooks fluctuation theorem and recovery of RNA folding free energies*, Nature (London) **437** (2005), 231.

[TC08] W. T. Coffey, Y. R. Kalmykov, S. V. Titov, and L. Cleary, *Smoluchowski equation approach for quantum Brownian motion in a tilted periodic potential*, Phys. Rev. E **78** (2008), 031114.

[TC09] W. T. Coffey, Y. P. Kalmykov, S. V. Titov, and L. Cleary, *Nonlinear noninertial response of a quantum Brownian particle in a tilted periodic potential to a strong ac force as applied to a point Josephson junction*, Phys. Rev. B **79** (2009), 054507.

[TH85] R. Kubo, M. Toda, and N. Hashitsume, *Statistical Physics II, Nonequilibrium Statistical Mechanics*, Springer-Verlag, Berlin, Germany, 1985.

[TH09a] M. Campisi, P. Talkner, and Peter Hänggi, *Fluctuation theorem for arbitrary open quantum systems*, Phys. Rev. Lett. **102** (2009), 210401.

[TH09b] _____, *Thermodynamics and fluctuation theorems for a strongly coupled open quantum system: An exactly solvable case*, J. Phys. A **42** (2009), 392002.

[Tho82] Sir W. Thomson, *Mathematical and physical papers*, Cambridge University Press, Cambridge, UK, 1882.

[TM07] W. T. Coffey, Yu P. Kalmykov, S. V. Titov, and B. P. Mulligan, *Semiclassical Klein-Kramers and Smoluchowski equations for the Brownian motion of a particle in an external potential*, J. Phys. A **40** (2007), F91.

[Tu04] L. Machura, M. Kostur, P. Hänggi, P. Talkner, and J. Łuczka, *Consistent description of quantum Brownian motors operating at strong friction*, Phys. Rev. E **70** (2004), 031107.

[uH06] L. Machura, M. Kostur, P. Talkner, J. Łuczka, and P. Hänggi, *Quantum diffusion in biased washboard potentials: Strong friction limit*, Phys. Rev. E **73** (2006), 031105.

[Uhl76] A. Uhlmann, *The transition probability in the state space of A^*-algebra*, Rep. Math. Phys. **9** (1976), 273.

[Ume62] H. Umegaki, *Conditional expectation in an operator algebra. IV. Entropy and information*, Kodai Math. Semin. Rep. **14** (1962), 59.

[Ved02] V. Vedral, *The role of relative entropy in quantum information theory*, Rev. Mod. Phys. **74** (2002), 197.

[VJ09] S. Vaikuntanathan and C. Jarzynski, *Dissipation and lag in irreversible processes*, Europhys. Lett. **87** (2009), 60005.

[vK92] N. G. van Kampen, *Stochastic processes in physics and chemistry*, Elsevier Science B. V., Amsterdam, The Netherlands, 1992.

[vN55] J. von Neumann, *Mathematical foundations of quantum mechanics*, Springer, Berlin, Germany, 1955.

[Vol65] V. Volterra, *Theory of functionals and of integral and integrodifferential equations*, McGraw-Hill, New York, USA, 1965.

Bibliography

[vZC03] R. van Zon and E. G. D. Cohen, *Stationary and transient work-fluctuation theorems for a dragged Brownian particle*, Phys. Rev. E **67** (2003), 046102.

[Wie86] F. W. Wiegel, *Introduction to path-integral methods in physics and polymer science*, World Scientific Publishing Co Pte Ltd., Singapore, 1986.

[Wig32] E. Wigner, *On the quantum correction for thermodynamic equilibrium*, Phys. Rev. **40** (1932), 749.

[Woo81] W. K. Wootters, *Statistical distance and Hilbert space*, Phys. Rev. D **23** (1981), 357.

[ZH95] C. Zerbe and P. Hänggi, *Brownian parametric quantum oscillator with dissipation*, Phys. Rev. E **52** (1995), 1533.

List of Figures

3.1 Equally spaced points in the sense of the statistical distance (3.2) (taken from [Woo81]). 35

3.2 Illustration of the definition of statistical length of a path through points with regions of uncertainty in a 3-dimensional probability space (taken from [Woo81]). 37

3.3 Upper plots: Exact mean energy (B.18) (dashed line) and linear approximation (3.50) (solid line) as a function of the time averaged Bures length (3.48) with $\beta\hbar\omega_0 = 1$, $M = 1$ and $\tau = 1$; Lower plots: corresponding parameterization of ω_t for arbitrary values of ω_1 50

3.4 Quantum speed limit time τ_{QSL}, Eq. (3.81), (solid line) and actual driving time τ (dashed line) for the linearly parameterized quantum harmonic oscillator (3.40) and (3.83) with $\hbar = \tau = 1$, $1/\beta = 0$ and $\omega_0 = 0.1$. 59

4.1 Cumulative heat distribution $\mathscr{P}_{\text{int}}(Q)$ (4.19) with $\gamma = \hbar\omega/20$, $\beta_A = 2$, $\beta_B = 1$, and $\tau = 10$ (left) and $\tau = 20$ (right). 67

4.2 Lower bounds for the entropy production $8/\pi^2 \mathscr{L}^2\left(\rho_\tau, \rho_\tau^{eq}\right)$ (solid line) and $\mathscr{D}^2\left(\rho_\tau, \rho_\tau^{eq}\right)$ (dashed line) as a function of the fidelity $F\left(\rho_\tau, \rho_\tau^{eq}\right)$. 74

4.3 Irreversible entropy production (4.55) (solid line) together with the lower bound (4.39) in lowest order expansion (dashed line) and including higher orders (4.38) (dotted line). 79

4.4 Maximal rate of irreversible entropy production, $\sigma_{\max}$, (4.50) as a function of Q^* (B.11) (left) and $\mathscr{D}$ (4.57) (right) 81

List of Figures

4.5 Electrode design. (a) Close-up view of the blade design with loading, taper, and experimental zone. (b) Sketch of assembled X-trap consisting of four blades. Compensation electrodes C1 and C2 are parallel to the trap axis. (Taken from [SSK08]) 82

4.6 (a) Contour plot of the potential in the $(x-y)$-plane in the experimental zone. (b) Cross-section trough potential along $x = y$-direction. (Taken from [SSK08]) 83

4.7 Variance and mean (inset) of the work for an oscillator with weak anharmonic corrections (solid line) (4.65) compared with those of the unperturbed oscillator (dashed line) (4.54) ($\omega_0 = 0.5$, $\tau = 1$, $\beta = 0.5$, $M = 1$, $\hbar = 1$ and $\sigma_\alpha = 0.025$). 88

4.8 Variance and mean (inset) of the work for a charged oscillator with weak electric noise (solid line) (4.65) compared with those of the unperturbed oscillator (dashed line) (4.54) ($\omega_0 = 0.5$, $\tau = 1$, $\beta = 0.5$, $M = 1$, $\hbar = 1$ and $\sigma_\alpha = 0.025$). 91

5.1 Defined quantum system and thermal surroundings 102

5.2 Phase space sketch illustrating the dynamics of the system; equilibrium (lower) and nonequilibrium (upper) path. 105

6.1 Exemplary smooth potential for typical escape problems. 124

6.2 Ratio of quantum Smoluchowski and classical escape rate, $\Gamma_{\mathrm{QSE}}/\Gamma_{\mathrm{cl}}$ (6.21), for the simple example $\omega_{\mathscr{A}} \equiv \omega_{\mathscr{B}}$ 127

6.3 Equivalent circuit for a resistively and capacitively shunted Josephson junction (RCSJ-model) 132

6.4 Stationary solution (6.49) for a circle-shaped junction with $C = 1.2pF$, $R = 0.37\Omega$, $I_c = 0.2mA$ and $\beta_c = 0.1$ and different temperature with: $\theta \simeq 1$ and $T = 0.973K$ (solid line); $\theta = 0.99$ and $T = 0.979K$ (large-dashed line); $\theta = 0.98$ and $T = 0.986K$ (dot-dashed line); $\theta = 0.97$ and $T = 0.993K$ (dashed line). 134

6.5 Temperature dependent I-V characteristics (solid line) (6.55) together with classically expected results (dashed line) (6.60) for circle shaped Josephson junctions with $C = 1.2pF$, $R = 0.37\Omega$, $I_c = 0.2mA$ and $\beta_c = 0.1$ and different values of i; left, i $= 0.985$; right, i $= 0.995$; 137

List of Figures

6.6 Temperature dependent I-V characteristics (solid line) (6.55) together with classically expected results (dashed line) (6.60) for circle shaped Josephson junctions with $C = 1.2pF$, $R = 0.37\Omega$, $I_c = 0.2mA$ and $\beta_c = 0.1$ and different values of i; left, $i = 1.005$; right, $i = 1.015$; 139

C.1 Particle trajectory which starts at position x_0 at time t_0 and passes trough $N+1$ intermediate time steps, $t_1, t_2, ..., t_N, t$ 162

List of Figures

I want morebooks!

Buy your books fast and straightforward online - at one of world's fastest growing online book stores! Environmentally sound due to Print-on-Demand technologies.

Buy your books online at
www.morebooks.shop

Kaufen Sie Ihre Bücher schnell und unkompliziert online – auf einer der am schnellsten wachsenden Buchhandelsplattformen weltweit! Dank Print-On-Demand umwelt- und ressourcenschonend produziert.

Bücher schneller online kaufen
www.morebooks.shop

KS OmniScriptum Publishing
Brivibas gatve 197
LV-1039 Riga, Latvia
Telefax: +371 686 204 55

info@omniscriptum.com
www.omniscriptum.com

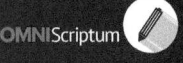

Printed by Books on Demand GmbH, Norderstedt / Germany